Geography Revision
for Leaving Certificate

PATRICK E. F. O'DWYER

GILL & MACMILLAN

Gill & Macmillan Ltd
Hume Avenue
Park West
Dublin 12
with associated companies throughout the world
www.gillmacmillan.ie

© Pat O'Dwyer 2005
0 7171 3846 1
Colour reproduction by Design Image
Print origination in Ireland by Carole Lynch

Acknowledgments
Maps and photographs reproduced by kind permission of Ordnance Survey Ireland
© Ordnance Survey Ireland and Government of Ireland.

The paper used in this book is made from the wood pulp of managed forests.
For every tree felled, at least one tree is planted,
thereby renewing natural resources.

Contents

OR

Elective 2: Patterns and Processes in the Human Environment (Topics 20–36)   167

SECTION 3: OPTIONS (CHOOSE ONE OPTION: HIGHER LEVEL STUDENTS ONLY)

Option 1: Global Interdependence (Topics 1–15)   204

OR

Option 2: Culture and Identity (Topics 16–30)

OR

Option 3: Geoecology (Topics 31–36)

OR

Option 4: The Atmosphere–Ocean Environment (Topics 37–42)

ATTENTION: LEAVING CERTIFICATE GEOGRAPHY STUDENTS

CORE

ALL STUDENTS MUST COVER

CORE TOPICS 1–21 INCLUSIVE (pages 1–130)

ELECTIVES

ALL STUDENTS MUST COVER *ONE* ELECTIVE

i.e. either

Elective 1: Topics 1–19 (pages 132–166)

or

Elective 2: Topics 20–36 (pages 167–202)

OPTIONS

HIGHER LEVEL STUDENTS ONLY

i.e. must cover *one* option

either

Option 1: Topics 1–15 (pages 204–242)

or

Option 2: Topics 16–30 (pages 243–272)

or

Option 3: Topics 31–36 (pages 273–292)

or

Option 4: Topics 37–42 (pages 293–311)

SECTION 1: CORE

UNIT 1:
PATTERNS AND PROCESSESS IN THE PHYSICAL ENVIRONMENT

UNIT 2:
REGIONAL GEOGRAPHY

All students must study **both** of these units
(pages 1–130)

CORE UNIT 1: PATTERNS AND PROCESSES IN THE PHYSICAL ENVIRONMENT

CORE TOPIC 1
STRUCTURE OF THE EARTH

THE CRUST

- The crust may be divided into the continents and the ocean floors.

  - **The Continents**
  - The continents are formed mostly of light, granite-like rocks. The continents have an average thickness of 45 km and they are up to 70 km thick under the mountain ranges.

  - **Ocean Floors**
  - The ocean floors are formed mostly of basalt, which is heavy. They have an average thickness of 8 km but may be as thin as 3 km in places.

THE MANTLE

- The continents, the ocean floors and the upper mantle form the Lithosphere. All the rocks in these areas are solid.
- The lower mantle consists of plastic-like rock that moves to form convection currents. The plates of the lithosphere move about on these slow-moving currents. The rock in the lower mantle is in a semi-liquid state because its temperature is very high.

THE CORE

The core is made up of nickel and iron. It is the hottest part of the Earth, where temperatures are greater than 4,000°C.

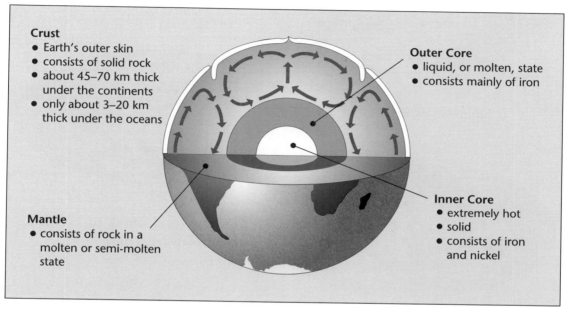

Crust
- Earth's outer skin
- consists of solid rock
- about 45–70 km thick under the continents
- only about 3–20 km thick under the oceans

Outer Core
- liquid, or molten, state
- consists mainly of iron

Inner Core
- extremely hot
- solid
- consists of iron and nickel

Mantle
- consists of rock in a molten or semi-molten state

▲ A section through the Earth.

CORE TOPIC 2

PLATE TECTONICS

- The Earth's crust is made up of **plates** that float on heavy, semi-molten rock and are moved around by convection currents beneath them.
- As the plates move around slowly, so do the continents and oceans that sit on top of them. This movement of the continents is known as **continental drift**.
- In places, these convection currents
  - drag the plates apart – these are plates in separation; or
  - push the plates together – these are plates in collision.
- High mountain ridges occur on the ocean floor in places where plates separate: for example, the Mid-Atlantic Ridge.
- **Fold mountains** are found in places where plates collide – e.g. the Armorican fold mountains of Munster and the Himalayas in the Indian subcontinent.

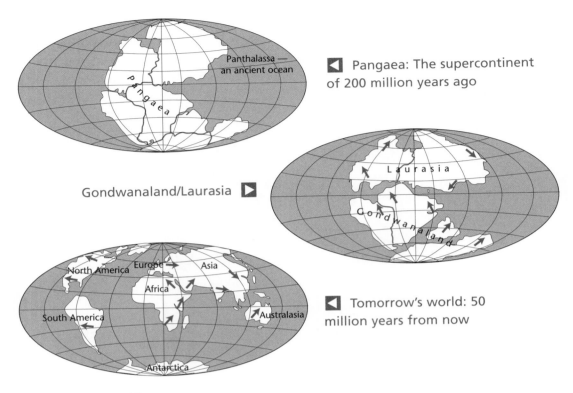

◀ Pangaea: The supercontinent of 200 million years ago

Gondwanaland/Laurasia ▶

◀ Tomorrow's world: 50 million years from now

CONTINENTAL DRIFT

- Over the past two billion years the continents of the earth have been moving about and constantly changing their position on the globe. This is the continental drift referred to above.
- During this process they have collided and separated many times. The last time they came together the continents collided to form a single, huge continent, called **Pangaea**.
- Pangaea was surrounded by a single ocean called **Panthalassa**.
- Then Pangaea initially split into two continents, **Gondwanaland** and **Laurasia**.
- Gondwanaland then broke apart forming Africa, Antarctica, South America and the Indian subcontinent. Laurasia split into Eurasia and North America.

◼ Proofs of Continental Drift

- Matching rocks found on continents that are thousands of miles apart
- Matching fossils that are found in precise locations where the continents were once joined together
- Matching edges of continents along the edges of the continental shelves, fitting together like a jigsaw puzzle.

THEORY OF SEA FLOOR SPREADING

- The **Theory of Sea Floor Spreading** suggests that ocean floors widen as new rock is formed along mid-ocean ridges where continents were split apart originally.

  ■ **Some Proofs of Sea Floor Spreading**
  - The existence of mid-ocean ridges
  - The varying ages of the sea floor. The age of the sea floor is youngest where new rock is formed along mid-ocean ridges, and oldest along continental edges.
  - Glacial deposits of similar types and ages are found in the areas where continents were attached.

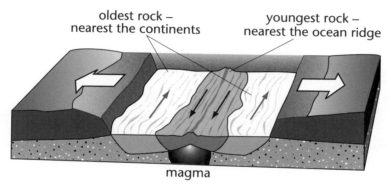

oldest rock –
nearest the continents

youngest rock –
nearest the ocean ridge

magma

▲ Sea floor spreading

CORE TOPIC 3
PLATE BOUNDARIES

New ocean floor is formed at boundaries of construction, where plates separate. To balance this so that the Earth's size does not increase, **old ocean floor** at the opposite side of the globe is sucked into the mantle and is destroyed. Most of this destruction occurs along the **Pacific Ring of Fire** along the edges of the Pacific Ocean.

■ **There are three types of plate boundary:**

Boundaries of construction
Boundaries of destruction
Passive boundaries.

BOUNDARIES OF CONSTRUCTION

New sea floor, new oceans and mid-ocean ridges form at boundaries of construction.

- A rising current of magma from the mantle splits the continent into two smaller continents.
- As these smaller continents move apart, sea water rushes in to fill the new valley.
- A mid-ocean ridge forms directly above the rising current of magma.
- Many volcanoes form along the mid-ocean ridge.
- Some, such as Iceland, may appear above the sea surface.

1. A new sea floor is created as the rising magma forms basalt when it meets the cold ocean water.

2. Ocean water spills in and fills the newly created valley to form a narrow sea

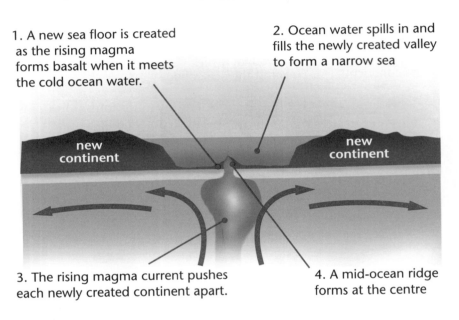

3. The rising magma current pushes each newly created continent apart.

4. A mid-ocean ridge forms at the centre

▲ Rising magma creates a mid-ocean ridge

BOUNDARIES OF DESTRUCTION

- Boundaries of destruction are places where old ocean floors sink into the mantle and are destroyed. This must happen so that the Earth does not get bigger because of increasing amounts of new sea-floor rock that are formed at mid-ocean ridges.
- Sea floors along the edges of the Pacific Ocean sink into the mantle. This process is called **subduction**. As they descend:
    - they melt, creating magma that forms curved lines of volcanoes, called volcanic arcs, at the surface;

- they become stuck, forming earthquakes along the line of the sinking plate;
- they create deep ocean trenches that form the deepest parts of the oceans, e.g. the Mariana Trench.

This zone, where the greatest amounts of subduction, volcanoes and earthquakes occur, is called the Pacific Ring of Fire.

■ **Subduction occurs in three types of location:**

1. Where two ocean plates collide

- One ocean plate sinks under the other.
- The sinking plate melts as it descends to form magma, which then rises to form a curved line of volcanoes called an island arc.

2. Where an ocean plate and a continent collide

- The heavier ocean floor sinks into the mantle.
- It buckles the land along the edge of the continent, forming fold mountains.
- It pushes, buckles and destroys islands and underwater plateaus and extinct volcanoes against the continent's edge, so making the continent wider.
- The descending plate melts to form magma, which then rises through the folded rock to form volcanoes within the fold mountains at the surface.

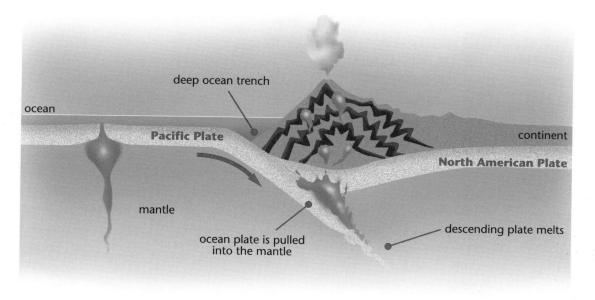

▲ Collision of Pacific and North American plates

3. Where two continents collide

- As two continents approach each other, the intervening ocean plate sinks under one or both of the continents.
- Finally all the sea floor and intervening islands and underwater plateaux are destroyed, and the continents collide to form high fold mountains such as the Himalayas.

PASSIVE BOUNDARIES

- Passive boundaries occur where plates slide past each other.
- Most of these boundaries occur along the edges of mid-ocean ridges.
- Some occur on dry land, such as the San Andreas Fault in California.
- Rock is neither created or destroyed at these boundaries.
- Many earthquakes occur along these fault lines.

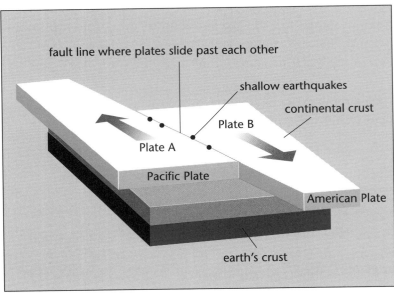

▲ The San Andreas Fault line in California

CORE TOPIC 4
VOLCANOES, EARTHQUAKES AND LANDFORMS

Volcanoes and earthquakes are located where plates separate and collide; but earthquakes also occur where plates slide past each other. The remainder of earthquakes occur along fault lines that are located away from plate boundaries.

Why do most earthquakes and volcanoes occur along the Pacific Ring of Fire?

- The Atlantic Ocean has a Mid-Atlantic Ridge that creates new land on its seabed. This also happens in the Indian Ocean and in the Pacific Ocean. To balance this, there are corresponding locations where old land is being recycled and sucked into the mantle. Most of this recycling, called subduction, occurs around the edges of the Pacific Ocean where ocean plates and continents meet.
- As the ocean plates sink into the mantle they melt to form volcanoes at the surface and they get stuck to form earthquakes.

■ **Sample Question and Answer**

Why has Plate Tectonics revolutionised our understanding of earthquakes and volcanoes? **Or** Explain why earthquakes and volcanoes regularly occur at similar locations.

Both earthquakes and volcanoes occur at constructive and destructive plate boundaries.

1. **Where two ocean plates collide is a destructive plate boundary**
- Ocean plates are heavy because they are formed of dense basalt rock and they are saturated with water. When two ocean plates meet one of them sinks under the other and slides into the Earth's mantle. As the ocean plate sinks into the mantle it sometimes gets stuck, and pressure is built up until it is suddenly released. This sudden release of energy causes earthquakes that occur near the surface of the seabed. These are called shallow earthquakes.
- As it sinks further into the mantle the moisture in the rock causes melting, which leads to intermediate earthquakes. The melting also creates magma that rises through the overlying rock of the other ocean plates and creates explosive volcanoes at the surface.
- These volcanoes create volcanic island arcs on the sea bed – such as Japan and the Philippine islands. When it sinks even further some of its minerals break up, causing deep earthquakes. All the earthquakes occur along the line of the sinking ocean plate. This line is called the **Benioff Zone**.

▼ Collision of two ocean plates

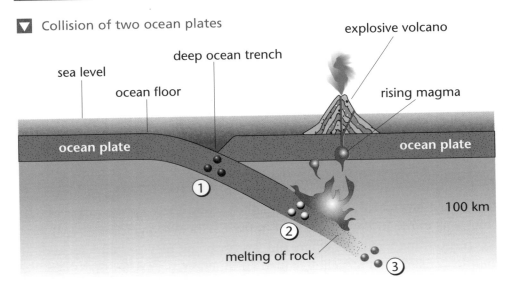

Example: Japan
1. shallow earthquakes
2. intermediate earthquakes
3. deep earthquakes.

2. **Where an ocean plate meets a continental plate, or where two continental plates collide, is a destructive plate boundary**

- The ocean floor is covered with thousands of metres of sediment that settled out of the sea water. As water-saturated ocean plates sink into the mantle at subduction zones under the continental plate, this thick sediment is scraped from the ocean floor. It is squeezed into layers of sedimentary rock and metamorphic rock that over millions of years is buckled and bent up into fold mountains.
- As the plate sinks it becomes stuck, just as before; strain is built up and finally the rock snaps, creating earthquakes.
- When the sinking plate reaches a depth of about 100 kilometres its water content causes surrounding rock to melt, creating magma and earthquakes. This magma rises through the buckled rock and some of it reaches the surface, creating highly explosive volcanoes. As the magma rushes to escape at the surface it also creates many earthquakes.
- Most of the world's earthquakes occur along the **Pacific Ring of Fire**, where ocean plates sink under continental plates: for example, in the **Rockies** in North America and in the **Andes** in South America.

When two continental plates approach each other the intervening ocean plate sinks under each continental plate, until the two plates collide.

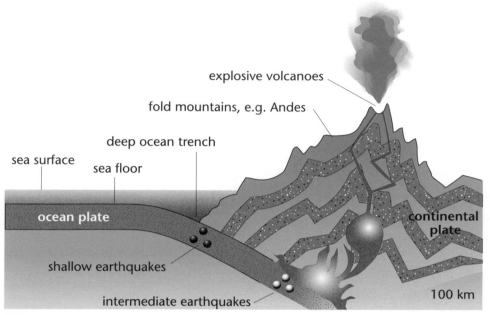

explosive volcanoes

fold mountains, e.g. Andes

deep ocean trench

sea surface

sea floor

ocean plate

continental plate

shallow earthquakes

intermediate earthquakes

100 km

▲ Subduction occurs where an ocean plate meets a continental plate

3. **Where a mid-ocean ridge creates new rock is a constructive plate boundary**

- Many earthquakes and volcanoes occur at mid-ocean ridges. As ocean plates separate, convection currents from the earth's mantle bring magma to the surface to fill the empty space on the sea floor. When this magma, at 800°C to 1,000°C, meets the cold ocean water it instantly becomes solid and forms new basalt rock.

- This new rock then splits, creating earthquakes, with one half attaching to each separating plate. Because this new rock is unable to withstand much strain it splits easily – just as new home-made bread would break much more easily than stale bread. So these earthquakes are generally small.

- Some places along mid-ocean ridges are hotter than others; they are called **hot spots**, where large volumes of magma pour out onto the sea bed. Generally this magma forms wide flows that build up into islands. Volcanoes also occur at these locations, such as **Iceland**. Many earthquakes and volcanoes occur along the **Mid-Atlantic Ridge**.

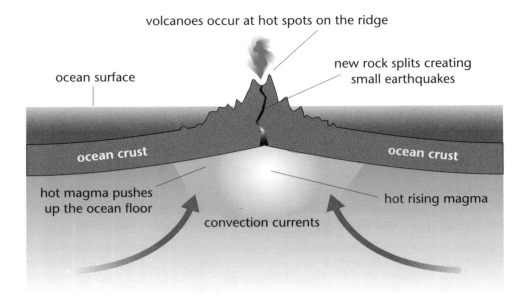

volcanoes occur at hot spots on the ridge

new rock splits creating
small earthquakes

ocean surface

ocean crust

ocean crust

hot magma pushes
up the ocean floor

hot rising magma

convection currents

▲ Convection currents pushing magma up through the ocean crust creating new sea floor

4. **Volcanoes do not occur where plates only slide past each other. These are passive boundaries**

- Only earthquakes occur where plates slide past each other. These are called transform faults. Most transform faults occur on the sea floor, but a major one occurs on land at the San Andreas fault in California. They are the means by which new rock that is created at mid-ocean ridges is carried to destructive boundaries, where plates sink into the mantle.

- While sliding past each other they sometimes become jammed. Strain is built up until the jamming point is unable to resist the pressure, and then the plates suddenly snap and jump forward. This sudden movement creates an earthquake – foreshocks and aftershocks. The strain is then transferred to the next jamming point, and the process repeats itself again and again.

- Because there is no subduction there is no melting of plates. So no magma exists to create volcanoes.

■ **The human cost of earthquakes is influenced by socio-economic factors**

1. **Countries of the Developed World**

   (a) Well-constructed buildings
   - Research into the way buildings move during earthquakes has led to new designs in building construction. Rich countries have the resources to invest in such research, because quality of life and safety matters are major issues during political elections. Buildings, especially tall ones, simultaneously bend and twist during earthquakes; therefore new structural engineering designs have been created so that these buildings will withstand the destructive power of reasonably severe earthquakes. New materials, such as special steel products, help to offset some damage.
   - Some buildings are built on 'roller' foundations, which allow the ground itself to move substantially while the building itself remains quite undisturbed, almost detached from the movement.

   (b) Education
   - Continuous education in schools on earthquake drill makes children, and later adults, aware of ways to remain reasonably safe during earthquakes. Children practise these exercises regularly in places that are especially prone, such as in Japan and California.
   - Fire prevention officers and specially trained civilians in key jobs are also trained in how to reduce the risk of serious injury during earthquakes.
   - Fire extinguishers are positioned in key locations such as kitchens, hallways and elsewhere, and automatic electric trip switches switch off current when some leakage occurs on electric lines.
   - Modern medical help and procedures also reduce death tolls in rich countries. Emergency plans that are practised under simulated conditions greatly help in regions of large populations, such as major industrialised cities of the developed world that are located close to earthquake zones.

   (c) Modern technology
   - Specialised equipment such as seismographs is strategically placed in regions of high earthquake risk. These instruments record foreshocks that indicate a major earthquake or volcanic eruption may be imminent. In such instances warnings are given to radio and television stations, so that people can be at least somewhat prepared for the unknown.
   - Tsunami warning stations have been set up for the Pacific Ocean region, where the most earthquake-prone cities in the world are located. This tsunami warning station is located in Hawaii, centrally located in the Pacific region and a state of the richest nation of the world, the USA.

Early warning can be relayed to areas at high risk of tsunami tidal waves that may result from an earthquake on the sea floor. When a warning is given, people in coastal regions may have time to evacuate coastal areas to avoid being killed.

2. **Developing Countries**

   (a) Poor buildings
   - Most residential buildings in developing countries are constructed of relatively loose materials such as mud bricks. They lack any structural fittings that are designed to resist earthquake damage and would reduce the risk of death or serious injury. During earthquakes, such buildings collapse on the occupants, killing most of them.
   - Many severe earthquakes have occurred in places such as India and Afghanistan, where as many as 50 to 60,000 people have been killed in a single town during a single earthquake.
   - Outside help is difficult to reach, as telecommunications are often poor or non-existent. In addition, many people live in large cities built close to major earthquake zones. Indonesia and India are located close to major fault lines that regularly cause earthquakes.

   (b) Lack of information systems
   - The lack of a tsunami warning system in regions such as the Indian Ocean has been directly influenced by a lack of resources being allocated to essential services. While many undeveloped countries have invested hugely in military equipment, they have neglected to invest in life-saving equipment and education for their coastal populations. This fact was directly responsible for the deaths of over 250,000 people during the tsunami disaster of 2004. In addition, the lack of coordinated information systems prevents local people from being made aware of impending disaster.

■ **How Volcanoes and their Effects may be Predicted**
- Types of material that form a volcano can indicate the power and explosive nature of a volcano.
- The dating of volcanic materials can create a timing pattern for eruptions.
- The distribution pattern of volcanic materials can suggest the area likely to be affected by a future eruption.
- Numerous small earthquakes near a volcano may suggest a larger eruption is about to occur.
- Changes in types and amounts of gases escaping from the sides or crater
- Changes in local ground water, temperature and composition.

■ Other Effects

1. **Nuee Ardente:** Generally volcanoes that occur at destructive boundaries are highly dangerous, because the magma contains a high proportion of silica and this prevents gases from escaping freely. Pressure builds up, and finally the whole mountain top is blasted into the sky. Such volcanic explosions give rise to Nuee Ardentes, which are clouds of poisonous gases, hot ash and rocks that rush down the volcano sides killing and destroying everything in their path.

2. **Lahar:** These are mud flows created by the sudden melting of ice by hot ash and lava on the sides of a volcano.

3. **Past volcanic eruptions:** The study of past volcanic eruptions in an area can help to prepare people for future eruptions.

Ice Ages

Thousands of metres of ice pressed down the land surface. When the ice sheets melted at the end of the Ice Age, the land gradually bounced back to its original level. This change still causes earthquakes from time to time.

Ancient Faults

Plates move along ancient faults that lie buried deep beneath the surface.

Earthquake Facts

- A seismologist is a person who studies earthquakes.
- A seismograph is an instrument that records and measures earthquakes.
- Earthquake strength is measured on the Richter Scale, from 0 upwards.
- The focus is the spot where an earthquake occurs.
- The epicentre is a spot on the Earth's surface directly above the focus.

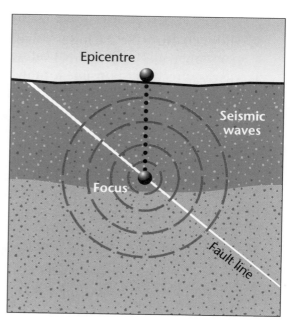

▲ A fault line through the focus of an earthquake

■ How Earthquakes and their Effects may be Predicted

Scientific instruments

Instruments are placed in regions that are liable to earthquakes. These regions include:

- fault lines
- the sides of active volcanoes.

The instruments measure any changes in the tilt or movement of the earth's surface, or a movement of rocks relative to one another.

Seismic Gaps

Places that have not had an earthquake for a long time but are bordered by areas of recent earthquake activity are the most likely spots for future earthquakes.

Dating Pattern

By establishing a pattern of past earthquake activity, one can predict the likelihood of a new earthquake.

■ Effects of Earthquakes

1. They can cause vertical and sideways displacement of parts of the Earth's crust.
2. They can cause the raising or lowering of parts of the sea floor.
3. They can cause the raising or lowering of coastal regions, as in Alaska in 1899, when some coastal rocks were lifted 15 m (50 feet).
4. They can cause landslides, as in the loess country of north China.
5. If shocks are experienced in densely populated and heavily built-up areas, the results can be disastrous. Some of the worst earthquakes include those of San Francisco in 1906 and Tokyo–Yokohama in 1923; the latter killed 100,000 people. In October 1989 San Francisco suffered another earthquake, which recorded 6.9 on the Richter scale and killed sixty-two people. In January 1994 Los Angeles suffered a severe earthquake centred on the San Fernando valley. Roads and buildings collapsed, and twenty-two people were killed.
6. Some gaping holes, cracks or subsidences are formed, railways and water pipes are cut and bridges collapse, causing massive structural damage and loss of life. Fires result from leaking gas pipes, while disease may occur in some areas from a combination of burst sewerage mains and high temperatures.

7. Where an earthquake affects the ocean floor, great waves, known as seismic waves or **tsunamis**, may spread outwards across the ocean at speeds of 500–800 km/h (300–500 miles per hour) and can cause great damage and a high death toll in coastal areas. Upon entering shallow coastal waters these destructive waves are slowed, and the water begins to pile up to heights that occasionally exceed 30 m (100 feet).

CORE TOPIC 5
FOLDING, ROCKS AND LANDFORMS

THE CALEDONIAN FOLDINGS

- The ancient ocean, the Iaepetus ocean, which lay between the Eurasian plate and the American plate, started to get smaller when these two plates moved towards each other.
- As they got close, small pieces of crust, called terranes, were squashed together to create the island of Ireland.
- At this stage Ireland was located about 30° south of the Equator, at about the same latitude where South Africa is today.
- As these terranes made contact with each other, the ones that formed the west and north-west of Ireland moved sideways, creating thrust faults that separate them today.
- Then the American and Eurasian plates collided about 400 million years ago to form one huge continent.
- The in-between ocean floor was subducted under both continents and the seafloor sediments were buckled up to form the sedimentary rocks of the Caledonian Fold Mountains.
- The Connemara and Wicklow mountains, the Scottish and Scandinavian Highlands and the Appalachians in North America were once part of this huge mountain range.

■ Rocks of the Caledonian Fold Mountains

- The subducted ocean floor melted under the mountains and then rose up through the buckled rock layers to form masses of magma.
- This magma cooled slowly to form **batholiths** of an igneous rock called **granite**. Granite has large crystals of **mica**, **feldspar** and **quartz**.
- The heat from the magma in the batholiths changed (metamorphosed) the sandstone into **quartzite**.

▶ Igneous and metamorphosed rock of the Wicklow mountains

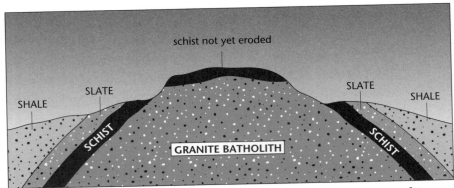

- Since then both the granite and the quartzite rocks have been exposed to form very different landscapes and landforms.
- Layers of shale that was heated under pressure changed (metamorphosed) to slate or schist.

■ Landforms of the Caledonian and Armorican Foldings

- The batholiths of granite now form **granite landscapes** of rounded hilltops, with blocks of granite in some places forming **tors** on top. These were formed when the overlying rock layers were eroded by weathering and erosion.
- The quartzite forms **quartzite landscapes** with very pointed sugarloaf peaks due to frost action. These include the Great and Little Sugar Loafs in Wicklow, the Twelve Bens in Connemara, Croagh Patrick in Mayo and Mount Errigal in Co. Donegal.
- Tors and granite landscapes also formed in Devon and Cornwall in southern England during the Armorican (Variscan) Foldings about 300 million years ago.
- Granite batholiths were formed in the folds of the sedimentary rocks in Devon and Cornwall because here the folding was intense, far greater than it was in Munster.

■ Formation of Ireland's Sedimentary Rocks

- Once the Caledonian Fold mountains with their sedimentary rocks, their metamorphic rocks and their igneous rocks had been formed, they came under severe erosion from weathering and other erosion processes.
- The mountains became worn down and the various rocks were broken up into particles that were washed down and deposited to form new sedimentary rocks of various types in a hot, desert environment that was prone to flash floods.

■ Conglomerates

- Conglomerates were formed near the mountains. Large pebbles of quartz, quartzite, feldspars, shales and slates were washed down gullies and deposited in alluvial fans and screes to form conglomerates.
- Millions of years later at the end of the last ice age, large moraines and eskers of boulders, gravels and sands were also cemented together to form conglomerates.

■ Sandstone: location 30° south of the Equator

- Large quantities of quartz grains from the weathered granite in the Caledonian mountains were washed into the lowlands during flash floods to form sheets of sand. Later these quartz grains were cemented together with an iron oxide cement, giving them a red colour.
- Some of the quartz grains were blown by the wind during long, dry spells and formed sand dunes. These sand dunes were also cemented together to form fine sandstone.
- All these sandstones are called Old Red Sandstone.
- They were formed in hot desert conditions.

■ Shale

- Shales and mudstones formed far from the mountains in delta and seabed areas because they are formed of the lightest particles, which are transported by river waters. They formed clays out of the weathered feldspars from the granite in the mountains.

■ Limestone: at the Equator (350 million years ago)

- Limestone was formed when Ireland was at the Equator.
- Limestones of many types were formed either in deep oceans or shallow seas. The limestones were formed from great thicknesses of broken shells, fossils, coral or lime mud. They were cemented by a lime mud.

■ Chalk: north of the Equator

- Chalk was formed from billions of shells called coccoliths, when Ireland was north of the Equator in a warm sea. It was a clear sea with little sediment being brought from land, so chalk is pure white.
- It is found in Antrim.

LANDFORMS OF SEDIMENTARY ROCKS

The Dartry–Cuilcagh Uplands and bedding planes

As Ireland was moving close to the Equator it experienced times when it was lowered below sea level to varying depths. **Alternate layers of sandstone, shale and limestone** were laid down on top of one another, each separated from the next by a **bedding plane**.

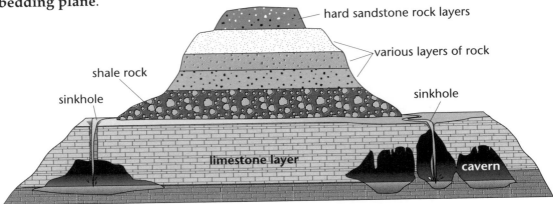

▲ Layers of sandstone, shale and limestone in the Dartry–Cuilcagh Uplands

- **■ Ridges and Valleys in Munster (300 million years ago)**
- During the Armorican (Variscan) foldings sedimentary rocks in Munster were folded to form ridges of sandstone and valleys of limestone.
- These ridges and valleys run west to east across Munster as do the rivers that flow in them.
- Limestone rock layers in the valley floors are younger than the sandstone layers on the ridges.
- Each rock layer is separated from the next by a **bedding plane**.

IRELAND'S BASALT ROCK AND ITS LANDFORMS (65 MILLION YEARS AGO)

Basalt is a fine-grained igneous rock.
It was formed when lava cooled quickly on the Earth's surface about 65 million years ago.
It forms six-sided columns and is found in the Giant's Causeway in Antrim.

- **■ Landforms**
- The Antrim Plateau is its largest landform.
- It forms volcanic plugs, sills, dykes and laccoliths.
- The Giant's Causeway in Co. Antrim.

Intrusive and extrusive landforms

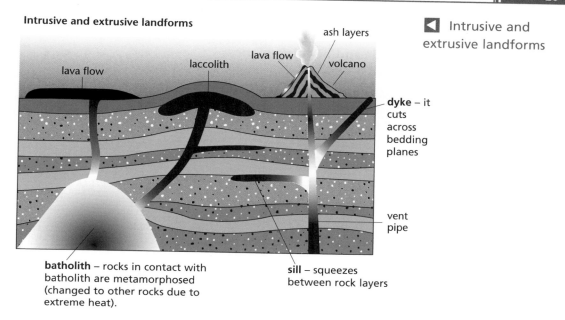

Intrusive and extrusive landforms

ash layers

lava flow

lava flow

laccolith

volcano

lava flow

dyke – it cuts across bedding planes

vent pipe

batholith – rocks in contact with batholith are metamorphosed (changed to other rocks due to extreme heat).

sill – squeezes between rock layers

LANDFORMS OF THE ALPINE FOLD MOVEMENT (37 MILLION YEARS AGO)

- The Alps are the youngest fold mountains. They were formed about 37 million years ago when the African Plate pushed into the Eurasian Plate.
- The Apennines were formed because the local sea floor sinks under the leg of Italy.

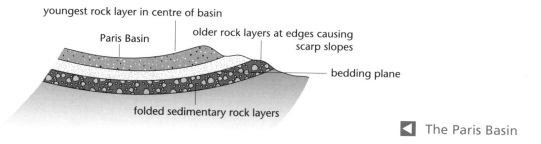

youngest rock layer in centre of basin

Paris Basin

older rock layers at edges causing scarp slopes

bedding plane

folded sedimentary rock layers

The Paris Basin

■ **The Paris Basin**
- Layers of sedimentary rock were slightly folded by the same earth movement that formed the Alps. The basin is dish-shaped, level at its centre and with steep slopes called **scarps** at its eastern edges.

■ **The Weald**
- A dome was created in the Weald in Southern England by folding. Weathering and erosion of the youngest rocks have exposed the older rocks in the centre.

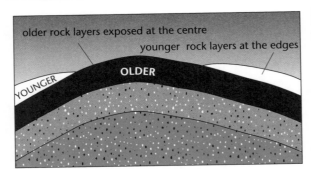

older rock layers exposed at the centre

younger rock layers at the edges

YOUNGER   OLDER

◀ The domed sedimentary rocks of the Weald

HOW PEOPLE INTERACT WITH ROCKS

■ In Ireland

Quarrying

Quarrying is the process whereby rock is blasted from quarry faces or excavated from the ground and prepared for the construction industry.

Sands and gravels are excavated from ridges, called eskers, and deltas of sands and gravels that were laid down by rivers that flowed at the end of the Ice Age. They are generally mixed with cement to make mortar and concrete.

Limestone and sandstone are blasted from quarry faces. They are then broken down into smaller particles of stone of various sizes called aggregates. These are used:

- to make concrete and concrete products, such as concrete blocks and roofing tiles
- for road surfacing
- as filling for passages and driveways
- limestone in powdered form, lime is used as a fertiliser.

Gypsum is quarried at Kingscourt in Co. Cavan. It is used to make plaster slabs for house construction.

Zinc is mined from limestone at Lisheen in Co. Tipperary, Galmoy in Co. Kilkenny and Navan in Co. Meath. It is used for weatherproofing metals for the construction industry, e.g. zinc-coated iron is called galvanised iron. It is used for roofing sheds and haybarns.

Marble is quarried in Connemara and Kilkenny. The term marble is widely used to mean any polished rock. However, pure marble forms from metamorphosed limestone.

■ **In Italy**

Marble is quarried at Carrara, in Tuscany. In its purest form it is white. It is used for flooring, wall tiles and fireplaces. Marble from these quarries was used by sculptors such as Michaelangelo Buonarroti during the Renaissance.

CORE TOPIC 6
AMERICA'S ACTIVE AND PASSIVE PLATE MARGINS

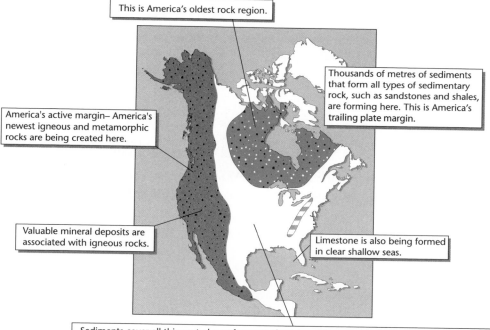

This is America's oldest rock region.

Thousands of metres of sediments that form all types of sedimentary rock, such as sandstones and shales, are forming here. This is America's trailing plate margin.

America's active margin– America's newest igneous and metamorphic rocks are being created here.

Valuable mineral deposits are associated with igneous rocks.

Limestone is also being formed in clear shallow seas.

Sediments cover all this central area from weathered rock from the Rocky and Appalachian mountains. Oil is found in this region because it is always found in recent sedimentary rocks with a high organic content.

▲ The American continent

AMERICA'S WEST COAST

- The Pacific Ocean floor is sinking under the American continent so this is an active plate margin.
- Rocks are buckled along this edge.
- Rising magma forms **batholiths** of igneous rocks.
- The heat from the batholiths forms marble, quartzite and slate.
- Most of America's newest igneous and metamorphic rocks are located along this edge.
- The Juan de Fuca plate sinks under the American plate.

AMERICA'S EAST COAST

- This is America's trailing plate margin.
- Sediments are being deposited by rivers on this continental edge.
- Thousands of metres of these sediments form all types of sedimentary rock, such as sandstones and shales.
- Limestone is also formed.

CORE TOPIC 7
LANDFORMS OF LIMESTONE ROCK

- Karst is a term used to describe extensive regions where bare limestone rock is exposed at the surface. Examples include:
  - The Burren
  - The Marble Arch Uplands.
- Limestone pavement is a large, bare, limestone region with grikes and clints formed by solution.
- Sinkholes and swallow holes: A sinkhole or swallow hole is an opening in the bed of a river where its water disappears underground.
- A cavern is a huge underground chamber formed by solution and erosion by rivers.
- Dripstone is calcite deposited in underground caverns. It forms stalactites, stalagmites, pillars and curtains.

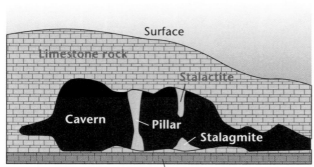

◀ Underground landforms in limestone rock

- Tower karst is a limestone landscape of steep, tower-like hills formed when all the surrounding caverns collapse. River deposits then create an alluvial flood plain.

CORE TOPIC 8

SURFACE PROCESSES

NB Students: Study all surface processes, pages 25 and 26.

PROCESSES OF MASS MOVEMENT

◼ **Soil creep**

This is the movement of soil particles downslope. It occurs due to the influences of:

• **gravity**, which pulls soil particles downslope.
• **solifluction**, which involves the swelling of some soil particles because they absorb ground water. This swelling causes neighbouring particles to move away from each other.
• **frost heave**, which involves the movement of soil particles by ice crystals that form under stones and moves them to the surface.

◼ **Gravity**

This pulls large rocks, boulders and soil downslope to create rockfalls and landslides.

◼ **Earth flows**

These occur when soil is saturated with water on gentle slopes.

◼ **Lahars or mud flows**

Lahars occur when enormous amounts of soil, rock, trees and other debris move rapidly downslope. They are triggered when large masses of ice melt owing to volcanic eruptions on ice-capped volcanic mountains.

◼ **Slumping**

This happens when cliff edges collapse; as they slip downwards there is a rotational movement of the falling material.

GLACIAL PROCESSES

◼ **Plucking**

Plucking involves water from melting ice trickling into cracks at the base of a glacier and then freezing. The glacier becomes attached to the rock under it, and then when the ice moves rock particles are plucked from the ground.

◼ **Abrasion**

This involves the use by the glacier of these plucked rocks to erode the base and sides of valleys, making them deeper. It also involves the scouring of lowland areas, leading to the removal of soil in some regions.

■ **Basal slip**

The sliding movement of a glacier over its rock floor.

■ **Freeze–thaw**

Freeze–thaw is what happens when by day melt-water seeps into cracks in rock and at night this water freezes and expands, breaking up the rock.

RIVER PROCESSES

■ **Hydraulic Action**

Hydraulic action is the breaking up of rock caused by the force of moving water.

■ **Corrasion or abrasion**

Corrasion is the use by a river of its load to erode the banks and bed of the river.

■ **Cavitation**

Cavitation occurs when bubbles of air collapse and form tiny shock waves against the outer bank of a river.

■ **Deposition**

Eroded material is dropped on the bed or flood plain of a river when the slope, the speed or the volume of a river is reduced.

■ **Attrition**

Fragments of stone are rounded and made smaller by hitting off each other.

■ **Slumping**

This is rotational movement of a collapsing river bank as it is undermined by a river.

COASTAL PROCESSES

■ **Abrasion**

Abrasion occurs when boulders, pebbles and sand are pounded by the waves against the coastline.

■ **Hydraulic Action**

The direct impact of strong waves on a coast.

■ **Compression**

Compression breaks up rock by air being squeezed in cracks and caves.

■ **Attrition**

Fragments of stone are rounded and made smaller by hitting off each other.

■ **Longshore Drift**

This is a zigzag movement of material along a shore. It builds up bars, spits and lagoons and leads to the development of salt marshes.

CORE TOPIC 9

SURFACE PROCESSES, PATTERNS AND LANDFORMS

NB Student Note:

Students should be able to **identify** all surface landforms by name or from a diagram.

STUDY ONLY ONE OF THE FOLLOWING IN DETAIL:
- Mass Movement, Processes, Patterns and Landforms: pages 27–29
- Glacial Processes, Patterns and Landforms: pages 30–36
- River Processes, Patterns and Landforms: pages 36–42
- Coastal Processes, Patterns and Landforms: pages 43–48

MASS MOVEMENT, PROCESSES, PATTERNS AND LANDFORMS

■ Factors that Influence Mass Movement

- Steepness of slope: the steeper the slope, the faster the movement.
- Type of material: loose material slips faster than compacted material.
- Water content: the higher the water content, the faster the movement.
- Vegetation cover: plant roots help to bind surface material to reduce movement.
- Earth movements: earthquakes shake and loosen material to aid movement.

■ Processes of Mass Movement

Soil creep

This is the movement of soil particles downslope. It occurs because of the influences of:
- **gravity,** which pulls soil particles downslope
- **solifluction,** which involves the swelling of some soil particles because they absorb ground water. This swelling causes neighbouring particles to move away from each other.
- **Frost heave,** which involves the movement of soil particles by ice crystals that form under stones and move them to the surface.

Gravity

This pulls large rocks, boulders and soil downslope to create rockfalls and landslides.

Earth flows

These occur when soil is saturated with water on gentle slopes.

Lahars or mud flows

Lahars occur when enormous amounts of soil, rock, trees and other debris move rapidly downslope. They are triggered when large masses of ice melt, owing to volcanic eruptions on ice-capped volcanic mountains.

■ Landforms of Mass Movement

Terracettes

These are parallel ridges of soil on a steep slope, much like long steps of a flight of stairs. They form because of:

(i)   Wet–dry periods: moisture increases the weight and volume of a soil causing expansion and the movement of the soil downhill under the pull of gravity. When the soil dries it contracts.

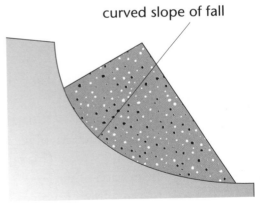

Terracettes

▲ Terracettes

(ii) Freeze–thaw: when a soil freezes its water particles expand and push up the soil at right angles to the slope. When the soil thaws the material slips downslope under the pull of gravity. This is the slowest form of mass movement.

Sloping Cliffs due to Slumping

Slumping is caused by undermining of a slope. When this happens, areas such as coastal cliffs and river cliffs collapse and slip and as they do there is a rotational movement of the falling material. Chalk cliffs at Garron Point in Co. Antrim were formed in this way.

curved slope of fall

Bog Bursts, Mud Flows and Earth Flows

These are moving masses of soil, stones, mud and water. If a saturated soil has a high content of fine soil particles, it is prone to sliding downslope. This may occur as a rapid or a slow movement.

▲ Slumping occurs along a curved slope

(i) Bog bursts

Blanket bogs that cover hill and mountain tops may flow after spells of continuous heavy rain; for example, this occurred at Derrybrien in Co. Galway. They can also occur when machinery is involved in peat removal just before, or during, wet spells.

(ii) Lahars

Lahars are particular types of mud flow that occur when hot lava or hot ash from a vent falls on great deposits of snow or ice on a high, snow-capped volcanic peak. The water from the melted snow or ice saturates the ground, which then rushes downhill rooting up soil, trees and other material on its journey. These flows may cover villages and towns at the foot of the volcano.

(iii) Earthflows

These may be caused by torrential rain saturating loose soil on a mountainside, which then flows downhill. In some instances such loose material may fall into a reservoir, causing vast amounts of reservoir water to pour over the dam and rush downhill, creating a mudflow as it goes.

■ Landslides and Rock Falls

Landslides

These occur when loose soil, stones and clay become loose and fall downhill owing to the pull of gravity. Water may or may not be an influencing factor. Landslides leave a concave scar on the slope where the material originated, and a convex mound of material where it ended.

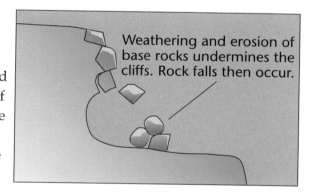

Weathering and erosion of base rocks undermines the cliffs. Rock falls then occur.

▲ Weathering and erosion of base rocks undermines cliffs, creating rock falls

Rock falls

Rock falls occur when loose boulders are set free on a slope and roll downhill. They may become loose for a number of reasons, including earthquakes or the erosion of surrounding soil or stones.

GLACIAL PROCESSES, PATTERNS AND LANDFORMS

■ Causes of Ice Ages

Ice ages occur because of three factors that happen to come together at the same time. These are:

- changes in the tilt of the Earth's axis every 41,000 years
- the orbit of the Earth changing from a circular path to a more elliptical one every 100,000 years
- the tendency of the Earth to 'wobble' every 23,000 years.

■ Glacial processes

Plucking

Plucking involves water from melting ice trickling into cracks at the base of a glacier and then freezing. The glacier becomes attached to the rock under it, and then when the ice moves rock particles are plucked from the ground.

Abrasion

This involves the use by the glacier of these plucked rocks to erode the base and sides of valleys, making them deeper. It also involves the scouring of lowland areas, leading to removal of soil in some regions.

Basal Slip

The sliding movement of a glacier over its rock floor.

Freeze–thaw is what happens when by day melt-water seeps into cracks in rock and then at night this water freezes and expands, breaking up the rock.

Definitions

Glacier: a river of ice.

Glaciated valley: a steep-sided and flat-floored valley (U-shaped) formed by the action of a glacier.

Crevasse: a long, narrow and deep crack in the surface of a glacier.

Fjord: a glaciated valley that has been drowned by sea water.

Erratic: a large boulder that was carried a long distance from its place of origin.

Outwash plain: a large, gently sloping area of sand and gravel that was dropped by streams that flowed from the front of an ice-sheet.

Overflow channel: a V-shaped valley cut by water that flowed from an ice-dammed lake.

Pyramidal peak: a peak formed when there were three or more cirques back to back and pointed by frost action (e.g. Carrauntoohil, Co. Kerry).

Arête: a knife-edged ridge formed where two cirques formed side by side.

LANDFORMS OF GLACIAL ACTION

Highland erosional landform

■ Landform: Cirque

Example: Devil's Punch Bowl in the Macgillicuddy's Reeks

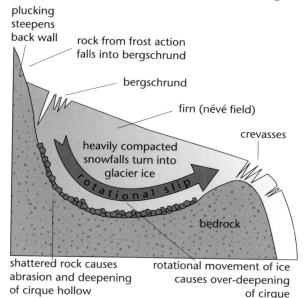

plucking steepens back wall

rock from frost action falls into bergschrund

bergschrund

firn (névé field)

crevasses

heavily compacted snowfalls turn into glacier ice

rotational slip

bedrock

shattered rock causes abrasion and deepening of cirque hollow

rotational movement of ice causes over-deepening of cirque

◀ The profile of a cirque lake

(margin text, rotated): Processes involved: abrasion, plucking, freeze–thaw.

Formation

Cirques are amphitheatre-shaped rock basins, variously known as cirques, cooms, cums or tarns.

They are formed when pre-glacial hollows are progressively enlarged on northern or north-eastern slopes. A patch of snow produces alternate thawing and freezing of the rocks around its edges, causing them to 'rot' or disintegrate. This process is called **snow-patch erosion**. As snowfall accumulates, large masses of ice form a **firn** or cirque glacier.

At this stage the ice moves downslope and pulls away from the headwall of the cirque, to which some ice remains attached. This gaping crack or crevasse is called the **bergschrund**. The back wall of the cirque maintains its steepness from melt-water, which seeps into cracks and, after alternate thawing and freezing, shatters the rock face. This action produces debris that falls down the bergschrund, freezes into the base of the ice field and acts as an abrasive. Ice movement pivots about a central point in the cirque. By **plucking** and **abrasion**, this action increases the depth of the hollow, which often contains a lake when the ice finally disappears.

■ **Landform: U-shaped Glaciated Valley**

Highland erosional landform
Example: Gap of Dunloe in Co. Kerry

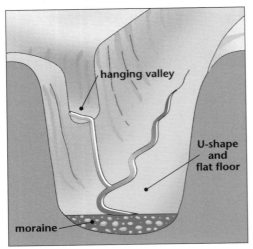

◀ Characteristics of a glaciated valley

(vertical side text) Processes involved: abrasion, plucking

Formation

When glaciers moved downslope through pre-glacial river valleys, they changed their V-shaped profile into wide, steep-sided U-shaped valleys. As the ice proceeded down the valley it used material that it **plucked** out from the valley floor to increase its erosive power. This material gathered on top of the glacier, within the glacier, and beneath the glacier. So, the gathered debris was used to increase vertical and lateral erosion in the valley. These processes of plucking and **abrasion** changed the pre-glacial, V-shaped valley into a U-shaped, glaciated valley. Most of our mountain valleys were glaciated (e.g. Cumeenduff Glen in Co. Kerry and Glenariff in Co. Antrim).

A glacier is a solid mass of ice that moves down a valley. Because of its solid nature it may have difficulty in passing through a valley that can vary in width from place to place. It overcomes this difficulty in a number of ways:
- **Compression** produces heat, and some ice melts, allowing the glacier to squeeze through, only to freeze again when the pressure is released.
- **Obstacles** in the glacier's path have a similar effect on the ice. Local melting on the upstream side allows the glacier to move over or around these obstacles as it moves downhill.
- **Friction** between the base of the glacier and the valley floor causes melting, producing a thin film of melt-water that acts as a lubricant, and so the glacier moves downslope.

Well-developed glaciated valleys are known as **glacial troughs.**

■ Landform: Ribbon and Pater Noster Lakes
Upland erosional landform
Examples: Lough Tay and Lough Dan in the Wicklow Mountains

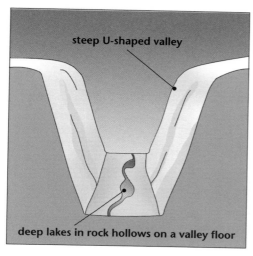

steep U-shaped valley

deep lakes in rock hollows on a valley floor

◀ Ribbon lakes are found on the floors of U-shaped valleys

Formation
Long stretches of glaciated valley floors may be over-deepened to create rock hollows. These hollows may have occurred because they were badly fractured owing to earlier earth movements, or were patches of softer rock that the ice more easily eroded. When these hollows were filled with water they may have formed isolated lakes on valley floors. We call these ribbon lakes. If they are found in a string that is joined by a river they are called **pater noster lakes**. Two processes mainly were involved in their formation. These are:

Plucking
Plucking involves water from melting ice trickling into cracks at the base of a glacier and then freezing. The glacier becomes attached to the rock under it and then when the ice moves, rock particles are plucked from the ground.

Abrasion
This involves the use by the glacier of these plucked rocks to erode the base and sides of valleys, making them deeper.

As glaciers moved through valleys, they passed over the fractured or soft rock patches. During cold spells or at night they may have stopped moving. At such times the liquid created by friction at the base of the glacier freezes and attaches the glacier to the bedrock. Once the glacier moves again, chunks of shattered or soft rock are plucked from the valley floor creating hollows. These hollows increase in size as the glacier continues to erode the valley. When all the ice has melted from the landscape, the hollows fill to form lakes.

■ Landform: Moraine

Upland and lowland depositional landforms
Example: Cumeenduff Glen in Co. Kerry

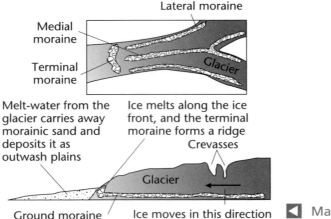

Lateral moraine

Medial moraine

Terminal moraine

Glacier

Melt-water from the glacier carries away morainic sand and deposits it as outwash plains

Ice melts along the ice front, and the terminal moraine forms a ridge

Crevasses

Glacier

Ground moraine   Ice moves in this direction   ◀ Main types of moraines

All rock material transported by a glacier, including boulder clay, is called **moraine**. Rock fragments range in size from large boulders to particles of dust.

Processes involved: transportation, deposition, melting of ice.

Formation

• *Lateral Moraine*

Long, sloping ridges of material left along valley sides after a glacier has melted are called **lateral moraine**. Freeze–thaw action is active on the **benches** (ridges) above glaciers, and angular rocks of all sizes fall onto the glacier edges below. This material accumulates to form lateral moraines. Vegetation may cover this material in time, and it may now be recognisable only by its lesser angle of slope compared to the valley walls or as a rocky sloping surface along valley sides.

• *Medial Moraine*

A medial moraine is formed from the material of two lateral moraines after a tributary glacier meets the main valley. These lateral moraines join and their material is carried down the valley by the main glacier. It is laid down as an uneven ridge of material along the centre of the main valley.

• *Terminal and Recessional Moraines*

When glaciers stopped for a long time during an interglacial or warm spell, they deposited an unsorted and crescent-shaped ridge of material across valleys and plains. These deposits have an uneven surface and are composed of moraine. In some instances they have caused moraine-dammed lakes to form by impeding drainage. In relation to upland areas, terminal moraines are found across the lower part of the mouth of a valley. Recessional moraines are found at higher levels up the valley.

■ Landform: Esker

Lowland depositional landform

Example: Eiscir Riada, near Clonmacnoise, Co. Westmeath; north and south of the River Brosna in counties Offaly and Westmeath

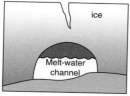
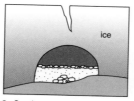
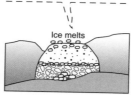
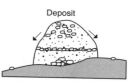

1. As ice melts, melt-water channels form under the ice.

2. Sand, gravel and boulders are deposited, depending on the speed of melt-water flow.

3. Melt-water channel fills with deposits as the ice melts.

4. After the ice has melted, esker slopes stabilise, leaving a ridge of sand, gravel and boulders.

▲ Esker formation

Processes involved: transportation, deposition

Formation

An **esker** or **os** is a long, low and winding ridge of sand and gravel that lies in the general direction of ice movement. Sections through eskers have revealed alternate layers of coarse and fine deposits, representing periods of rapid and slow ice melting, respectively. They represent the beds of former streams flowing in and under ice sheets. Changes in meltwater routes sometimes led to a section of tunnel being abandoned by the main stream flow. It would then silt up with sand and gravel; when the ice ultimately disappeared, the tunnel-fill would emerge as an esker, a ridge running across the country for several kilometres and bearing no relation to the local topography.

The surrounding landscape may have a boulder clay covering, giving rise to rich farmland that often stands in stark contrast to the sandy soils of an esker, which may display a poor-quality grass surface or coarse grasses and scrub. The esker was formed as the ice retreated rapidly.

■ Landform: Drumlin

Lowland depositional landform

Example: Hills in Co. Cavan and islands in Clew Bay in Co. Mayo

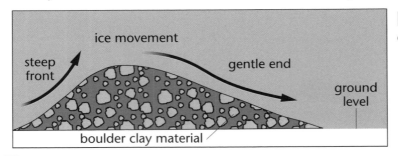

◀ Drumlins are oval-shaped low hills

Processes involved: transportation, deposition, ice-action

Formation

Drumlins are composed of unstratified (not in layers) ground moraine of rocks, pebbles, gravel, sand and clay all mixed up together. They represent the ground moraines of ice sheets that were stationary for some time. The materials that make up drumlins came from rocks that were plucked from the bedrock over which the ice passed and broken down into gravel and sand. Drumlins range in size from just small mounds to hills over one kilometre long and up to one hundred metres in height.

Drumlins were deposited as heavily laden ice sheets passed over a lowland landscape. They usually occur in clusters or swarms to form 'basket of eggs scenery'. Drumlins are generally rounded, oval-shaped or egg-shaped. Their long axis lies in the direction of the ice movement. The steeper end represents the direction from which the ice came.

Sometimes drumlins prevent local drainage and lead to marsh areas or turloughs between the hills.

OR

RIVER PROCESSES, PATTERNS AND LANDFORMS

■ River Patterns

A **basin** is the area drained by a river. The pattern of drainage in a basin may be:

- **dendritic**, when the tributaries form a pattern like the branches of a tree
- **trellised**, when tributaries run parallel to each other towards the main course and meet the main river at right angles
- **radial**, when streams flow downhill, radiating from a central hilltop or mountaintop.

■ River Processes

- **Hydraulic action** is the breaking up of rock caused by the force of moving water.
- **Corrasion** is the use by a river of its load to erode the banks and bed of the river.
- **Cavitation** occurs when bubbles of air collapse and form tiny shock waves against the outer bank of a river.
- **Deposition** material is dropped on the bed or flood plain of a river when the slope, the speed or the volume of a river is reduced.

Definitions

Source: the place where a river begins.

Tributary: a river that joins a larger one.

Confluence: the place where rivers join.

Mouth: the place where a river enters a sea or lake.

Estuary: that part of a river's course that is tidal.

Basin: the entire area drained by a river and its tributaries.

Watershed: the high ground that separates one river basin from another.

Landforms of River Action

Landform: Waterfalls

Landform of erosion; upper course

Examples: Asleagh Falls on the Erriff River; Torc Waterfall in Killarney

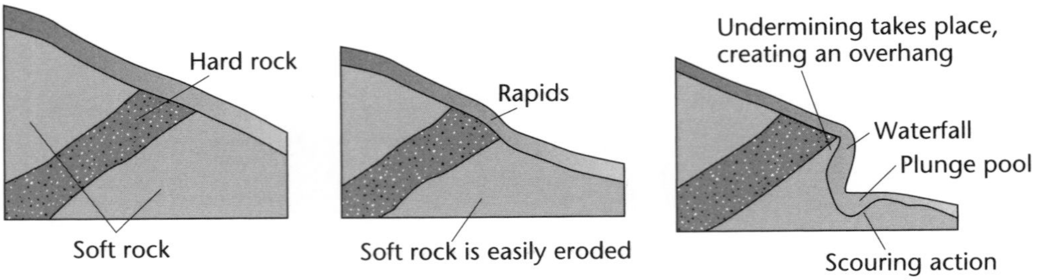

Waterfall formation

Processes involved in formation: hydraulic action, corrasion, eddying, solution, rejuvenation

Formation

- When waterfalls occur in the upper course of a river, their presence usually results from a bar of hard rock lying across the valley of the river. If this band of rock is horizontal or slightly inclined, a vertical fall in the river results.
- The scouring action of the falling water and the river's load at the base of the fall cut into the underlying soft rock, creating a **plunge pool**.
- Two processes are involved in creating the plunge pool: **hydraulic action** and **corrasion.**
  1. Hydraulic action is caused by the force of the falling water; by rushing into cracks, the water can help to break up solid rock. Corrasion is the use by the river of its load to erode the side and bed of the river. At the base of a waterfall, turbulent water and **eddying** by the river and its load erode the bed to form a plunge pool. Undermining causes an overhanging ledge of hard rock, pieces of which break off and collect at the base of the waterfall. As the fall recedes upstream, a steep-sided channel is created downstream of the falls. This feature is called a **gorge**.
  2. If a waterfall appears elsewhere in the course of a river, it may be the result of **rejuvenation**. This may be caused by a fall in sea level, a local uplift of land, which causes a steeper slope and a greater river speed and so renews downcutting or vertical erosion. This produces a new curve or profile of erosion that intersects with the old curve at the **knickpoint**. **Rivers in Co. Donegal have been rejuvenated,** and many flow over waterfalls before they enter the sea.

■ **Landform: Flood Plain**

Landform of erosion and deposition; middle and lower course
Example: River Shannon in the midlands; Blackwater Valley near Fermoy

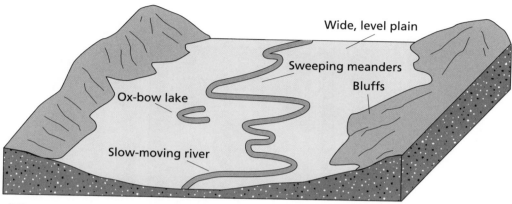

Wide, level plain
Sweeping meanders
Bluffs
Ox-bow lake
Slow-moving river

▲ Flood plain with sweeping meanders

Processes involved in formation: undercutting, divagation, deposition

Formation

When meanders migrate downstream, they swing to and fro across a valley. As a result the river swings from side to side. As the water flows around a bend, it erodes most strongly on the outside, forming a river cliff. **Undercutting** of the bank takes place. There is little erosion on the inside of a bend, but deposition often occurs, causing a gravel beach or slip-off slope. The valley has been straightened at this stage, with interlocking spurs removed by the lateral erosion of the meanders (**divagation**), and a level stretch of land is created on both sides of the river. This is called a **flood plain**.

During its upper and middle stages, a river flows quickly and is able to transport a large amount of material, called its **load**, by the processes of solution, suspension and saltation. However, in the lower stages the speed of a river is reduced, because of the more gentle slope of the valley floor. At this stage the river is able to carry only the smallest particles of silt and clay, collectively called **alluvium**.

A flood plain is a wide and flat valley floor that is often subjected to flooding during times of heavy rain. When this occurs, the river spreads across the flat flood plain and deposits a thin layer of alluvium. Alluvium is fine material consisting of silt and clay particles and is rich in mineral matter, transported by a river and **deposited** at places along the flood plain. This deposit enriches the soil and leads to the creation of fertile farmland.

■ **Landform: Ox-Bow Lake**

Feature of erosion and/or deposition; middle and lower course
Example: River Shannon at Leitrim Town

A neck of land separates two concave banks where erosion is active

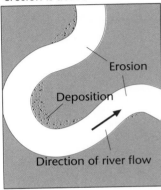

Neck is ultimately cut through; this may be accelerated by river flooding

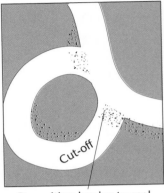

Deposition seals the cut-off, which becomes an ox-bow lake

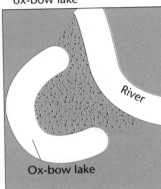

Deposition begins to seal up the ends of the cut-off

▲ Ox-Bow lake formation

Formation

As meanders move downstream, erosion of the outside bank leads to the formation of a loop in the river's course, enclosing a peninsula of land with a narrow neck. Three main processes of river erosion act together to create the ox-bow lake. **Hydraulic action** is caused by the force of the moving water. By rushing into cracks and by direct contact with the river banks, it can help to break up solid rock and undermine the banks. **Corrasion** is the use by the river of its load to erode, in this case, the river bank. Along the side of the outer bank turbulent water and **eddying** (swirling movement) by the river and its load create a river cliff. Erosion also occurs when bubbles of air collapse and form shock waves against the outer bank. Loose clays, sands and gravels are quickly worn away by this type of process.

Finally, during a period of flood the river cuts through this neck and continues on a straight and easier route, leaving the river loop to one side. Deposition occurs at both ends of this loop to form an ox-bow lake.

After a long time these ox-bow lakes become filled with silt from flood water, and finally they dry up. At that stage they are called **meander scars** or **mort lakes**.

Processes involved in formation: hydraulic action, corrasion, cavitation, deposition

■ Landform: Delta

Landform of deposition; lower course
Example: Roughty River in Co. Kerry; the Cloghoge river delta in Lough Tay in Co. Wicklow

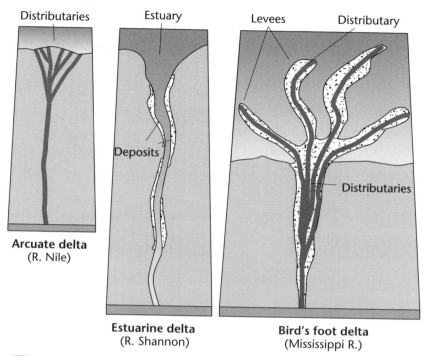

Arcuate delta
(R. Nile)

Estuarine delta
(R. Shannon)

Bird's foot delta
(Mississippi R.)

▲ Types of delta

Formation

The materials deposited in a delta are classified into three categories:

1. Fine particles are carried out to sea and are deposited in advance of the main delta. These are the **bottom-set beds**.
2. Coarser materials form inclined layers over the bottom-set beds and gradually build out, each one in front of and above the previous ones, causing the delta to advance seawards. These are the **fore-set beds**.
3. On the landward margins of the delta, fine particles of clays, silts and muds are laid down, continuous with the river's flood plain. These are the **top-set beds**.

When a river carries a heavy load into an area of calm water, such as an enclosed or sheltered sea area or a lake, it deposits material at its mouth. This material builds up in layers called **beds** to form islands, which grow and eventually cause the estuary to split up into many smaller streams, called **distributaries**. Should this occur in a lake, it is called a **lacustrine**

Processes involved: deposition

delta (e.g. Glendalough, Co. Wicklow). If it occurs at a coast, it is called a **marine delta** (e.g. the Roughty River in Kenmare Bay, Co. Kerry). The material that builds up to form the delta is composed of alternate layers of coarse and fine deposits, which reflect times of high and low water levels, respectively, in the river. Mountain streams flowing into glaciated valleys often build deltas in ribbon lakes. This causes a filling-in of the lake, reducing its length over time or dividing ribbon lakes into separate lakes.

■ **Landform: Levees**

Landforms of deposition; lower course

Example: Mulkear River in County Limerick

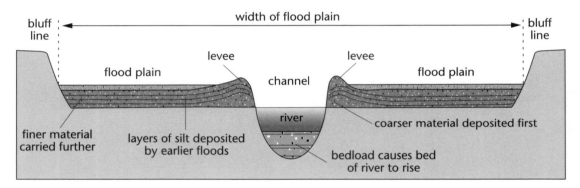

◢ Levee formation

Formation

Levees are high banks along a river's edge in its lower course. They form naturally or can be man-made to retain flood waters and prevent flooding of the surrounding low flood plain. They generally form along the edges of silt-laden rivers as the rivers slowly wind their way across flat flood plains to the sea. The best-known levees are found along the Yangtze Kiang river in China, on which many villages are built.

As sediment-laden flood water flows out of its completely submerged channel during a flood, the depth, force and turbulence of the water decrease sharply at the channel margins. This decrease results in a sudden dropping of the coarser materials, usually fine sand or silt, along the edges of the channel, building up a levee.

As a levee increases in height it will eventually retain all floodwaters and prevent the river from overflowing its banks onto the flood plain.

Man-made levees are also constructed to prevent flooding and damage to villages and towns, especially where large populations live on floodplains.

Process involved: deposition

OR

COASTAL PROCESSES, PATTERNS AND LANDFORMS

■ Coastal Processes

- **Abrasion** occurs when boulders, pebbles and sand are pounded by the waves against the coastline.
- **Hydraulic action** is the direct impact of strong waves on a coast.
- **Compression** is the breaking of rock by air compressed in cracks and caves.
- **Attrition** happens when fragments of stone are rounded by hitting off each other.
- **Longshore drift** is a zigzag movement of material along a shore. It builds up bars, spits and lagoons and leads to the development of salt marshes.

Definitions

Wave: Wind causes water particles on the surface of the sea to move in a circular motion and form a wave shape. This disturbance is transmitted to neighbouring particles, and so the wave shape moves forward (not the actual water).

Swash: water that rushes up a beach following the breaking of a wave.

Backwash: the return of the water down the beach.

Longshore drift: the movement of material along the shore.

Load: mud, sand and shingle carried along the shore by the sea.

LANDFORMS OF COASTAL ACTION

■ Landform: Sea-Stack

Landform of erosion

Example: Kilkee coast in Co. Clare

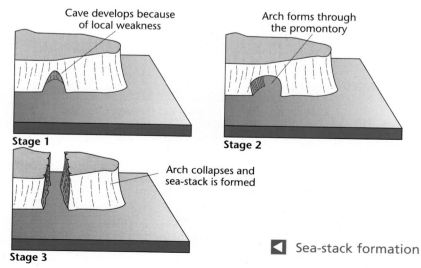

Cave develops because of local weakness

Arch forms through the promontory

Stage 1

Stage 2

Arch collapses and sea-stack is formed

Stage 3

◀ Sea-stack formation

Formation

Hydraulic Action

The direct impact of strong waves on a coast has a shattering effect as it pounds the rocks. Strong waves breaking against the base of a cliff force rocks apart, creating a cave.

Compression

Air filters into joints, cracks and bedding planes in a cave. This air is trapped as incoming waves lash against the coast, and the trapped air is compressed until its pressure is equal to that exerted by the incoming wave. When the wave retreats, the resultant expansion of the compressed air has an explosive effect, enlarging fissures and shattering the rock face. Caves are formed and enlarged in this way to form sea-arches.

Corrasion and Abrasion

When boulders, pebbles and sand are pounded by waves against the coastline, fragments of rock are broken off. The amount of corrasion is dependent on the ability of the waves to pick up rock fragments from the shore. So corrasion is most active during storms and at high tide, when incoming waves throw water and rock material against the coastline, eroding the sea-arch until it collapses and leaving a sea-stack isolated from the coastline.

■ **Landform: Cliff**

Landform of erosion
Example: Cliffs of Moher, Co. Clare

Formation

As destructive waves lash against a shoreline, a wedge of material is eroded to form a notch. As the notch is enlarged, a steep cliff is formed. Undercutting of this cliff face occurs as the waves continue to strike at the base of the cliff. In time the overhang collapses and

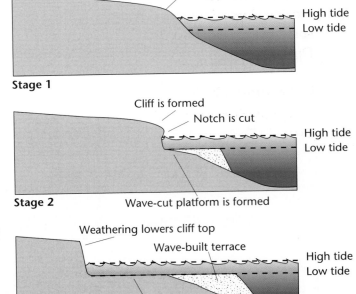

▲ Cliff formation

the cliff retreats. The following processes of erosion are involved in cliff formation.

Hydraulic Action

The direct impact of strong waves on a coast has a shattering effect as it pounds the rocks. Strong waves breaking against the base of a cliff force rocks apart, making them more susceptible to erosion. Cliffs of boulder clay are particularly affected, as loosened soil and rocks are washed away.

Compression

Air filters into joints, cracks and bedding planes in a cave. This air is trapped as incoming waves lash against the coast, and the trapped air is compressed until its pressure is equal to that exerted by the incoming wave. When the wave retreats, the resultant expansion of the compressed air has an explosive effect, enlarging fissures and shattering the rock face.

Corrasion and Abrasion

When boulders, pebbles and sand are pounded by waves against the coastline, fragments of rock are broken off, and undercutting of the cliff takes place.

As a cliff face retreats, a wave-cut platform is formed at its base. This is a level stretch of rock, often exposed at low tide, with occasional pools of water and patches of seaweed on its surface and displaying a boulder beach in the backshore. Generally the wider a wave-cut platform, the less the erosive power of waves, as shallow water reduces wave action, so the rate of coastal erosion slows down. Wave-cut platforms occur above the present sea level in some parts of the country, such as at Black Head, Co. Clare, and Annalong, Co. Down. These were formed when the sea was at a higher level than it is today.

■ **Landform: Beach**

Landform of deposition

Example: Tramore Beach in Co. Waterford

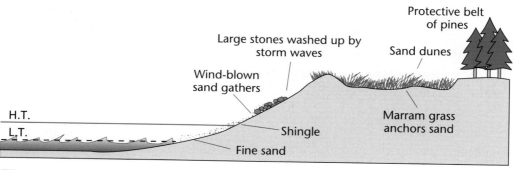

▲ Composition of a beach

Formation

The term **beach** is applied to the accumulation of material between low-tide level and the highest point reached by storm waves. This material usually consists of stones, pebbles, shingle and sand, all of which were deposited by constructive wave action.

An ideal beach profile has two main parts:

(a) the **backshore**, which is composed of rounded rocks and stones and broken shells, pieces of driftwood, and litter thrown up by storm waves. This part of the beach has a steep slope and is reached by the sea during the highest tides or during storms

(b) the **foreshore**, which is composed of sand and small shell particles, has a gentle gradient, and is covered by the sea regularly each day.

Longshore Drift

Longshore drift is the movement of material (sand and shingle) along a shore. When waves break obliquely onto a beach, pebbles and sand are moved up the beach at the same angle as the waves by the swash. The backwash drags the material down the beach at right angles to the coast, only to meet another incoming wave, and the process is repeated. In this way materials are moved along the shore in a zigzag pattern.

Gradually there is a net gain of material on the shore, creating a beach. Sometimes beaches may be composed of large stones and pebbles. This is called a boulder beach. Some beaches may have ridges of sand separated from each other by depressions called runnels.

■ **Landform: Lagoon**

Landform of deposition
Example: Lady's Island Lake in Co. Wexford

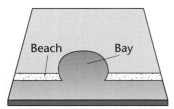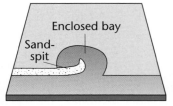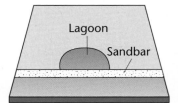

▲ Lagoon formation: method 1

Formation

Sandspit

A spit is formed when material is piled up in line-like form, but with one end attached to the land and the other projecting into the open sea, generally across the mouth of a river. Spits generally develop at places where longshore drift is interrupted and where the coastline undergoes a sharp change of direction, such as at river mouths, estuaries and bays, or between an island and the shore. Deposits gradually build up to form a projecting ridge of beach material. This growth continues for as long as the amount of beach material being deposited is greater than the amount that is removed.

Sandbar

A sandbar may form as a result of the growth of a sandspit across a bay. When a bay is cut off from the sea, a **lagoon** is formed. Sometimes where tidal or river scouring takes place, a bar may be prevented from completely sealing off the bay, as is the case along the coast near Wicklow. This type of sandbar is called a **baymouth bar.**

Stages of a Lagoon

A lagoon is formed when waves build a bar above water and across a bay or parallel to the coastline. Waves wash sand into the lagoon, and rivers and winds carry sediment into it. The lagoon becomes a marsh. Finally, the work of waves, rivers and winds turns the marsh into an area of sand dunes.

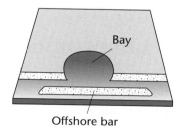

Bay

Offshore bar

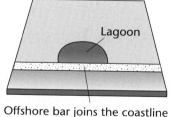

Lagoon

Offshore bar joins the coastline to form a baymouth bar

▲ Lagoon formation: method 2

Offshore Bar

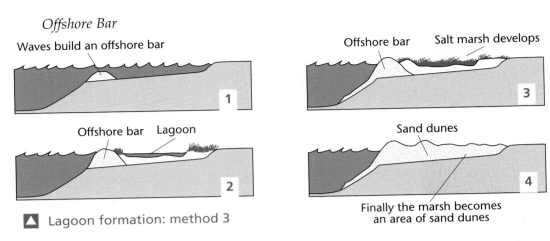

Lagoon formation: method 3

Processes involved:
deposition, longshore drift

An offshore bar is a ridge of sand lying parallel to a shore and some distance out to sea. On gently sloping coasts, **breakers** (large ocean waves) break, dig up the sea bed, and throw the loose material forward to form a ridge of sand. Once the ridge is formed, the bar increases in height by constructive wave action.

These ridges are pushed along in front of the waves until finally they may lie across a bay to form a baymouth sandbar. Lady's Island Lake and Tacumshin Lake in Co. Wexford were originally bays of the sea before being cut off by such a sandbar.

CORE TOPIC 10
THE PROCESS OF ISOSTASY AND LANDFORMS

The Earth's crust is made up of rocks of differing densities.
- The continents are composed of light rocks, called **sial**.
- The ocean floors are made up of dense rocks, called **sima**. A continent 'floats' on a layer of sima that runs under the continents and along the floors of the oceans.
- When erosion of continents occurs sediment is deposited in lowlands, increasing weight in this area – so they are pressed down.
- This action also reduces weight in mountain regions, so the mountain regions float higher on the sima layer. Together these actions cause a levelling of the landscape. This process is called **isostasy**.

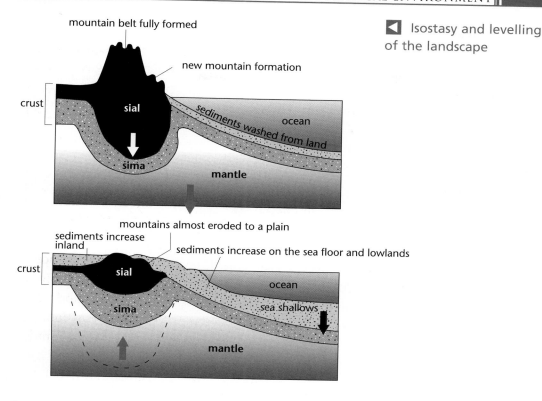

◄ Isostasy and levelling of the landscape

LANDFORMS CAUSED BY CHANGES IN SEA LEVEL

- The process of **isostasy** causes raising and lowering of land relative to the level of the sea. This can lead to some landforms being drowned, or partially drowned, by the sea.
- On other occasions landforms may be raised above sea level, when they were once either at sea level or below it. Earth movements can also create these landforms.

■ Emerged Coastal Features

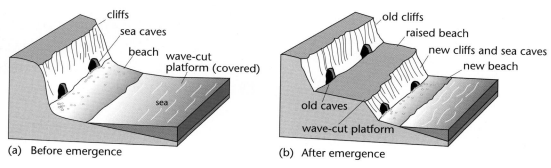

(a) Before emergence

(b) After emergence

▲ Raised beach and wave-cut platform

Raised Beaches and Wave-Cut Platforms

- The sea level, relative to the land, changes over time. If the level of the sea falls or the land rises, then coastal features such as beaches or wave-cut platforms may now be well above sea level.
- Step-like terraces may also form when sea levels change.

■ Submerged Coastal Features

Rias

- Rias are submerged river valleys. They occur in south-west Ireland; Dingle Bay and Bantry Bay are rias.
- As the American plate moved away from the Eurasian plate the west coast of Ireland lost its support, tilted seawards and was drowned by the sea. This process created the rias of the south-west.

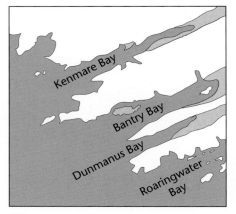

▲ Rias in the south-west of Ireland

Fjords

- During the last ice age, glaciers carved deep, U-shaped valleys in coastal mountain ranges. When the Ice Age ended the water stored in the ice flowed back into the sea, causing the sea level to rise.
- The rising sea filled the deep valleys, forming fjords. Killary harbour is Ireland's best fjord example. Norway has many fjords.

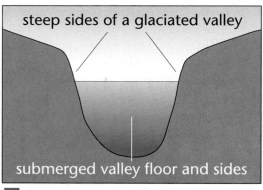

▲ Features of a fjord

ADJUSTING TO BASE LEVEL

- When earth movements raise land, the rivers in that region will erode to create a new curve or profile. Rivers begin cutting their new curve or profile from their estuaries upstream.
- This process is called **rejuvenation**.
- Waterfalls or rapids on a river near to its estuary indicate that uplift has recently taken place. In Co. Donegal in Ireland, the rejuvenation was caused by loss of weight on the landscape when the Ice Age ended.

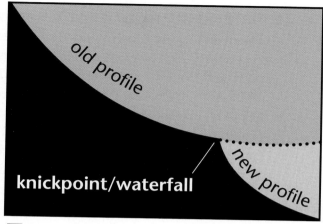

old profile

knickpoint/waterfall

new profile

▲ Waterfall (old and new profiles)

- **Waterfalls** occur where the old profiles and new profiles meet. This is called the **knickpoint**.

CYCLE OF LANDSCAPE EVOLUTION

- As new, level landscapes are exposed to weathering and erosion, they come under attack from rivers, wind, rain, frost and mass movement.
- The rivers open up channels, and the other processes combine to divide the original level land into separate ridges and valleys.
- These ridges gradually get worn down until they are just barely visible as raised land separating valleys that are in their lower stages of development.
- These almost perfectly flat landscapes are called **peneplains**. Example: South Cork.

CORE TOPIC 11

PEOPLE'S INTERACTION WITH SURFACE PROCESSES

Students should study only ONE of the following:
- Mass movement processes: page 52
- River processes: pages 52–53
- Coastal processes: page 54

MASS MOVEMENT

■ **The Effect of Overgrazing on Mass-movement Processes**

- The overstocking of land may directly lead to soil erosion. When vegetation, such as grass or heather, is overgrazed, the soil is no longer protected from the direct impact of raindrops. Then raindrops strike the soil and loosen it, causing it to be washed downhill.
- In places like southern Italy, slopes are steep and rain falls in isolated, heavy downpours during hot weather. This can lead to **mudflows**.
- **Constant erosion by winter rains** over long periods on steep slopes removes most of the soil and washes it into river channels, creating flooding.
- Since the 1990s Irish mountainsides such as the Galtee and Mweelrea mountains have been subjected to increased mass movement, as overstocking of sheep has led to soil erosion.

■ **The Effect of Overcropping on Mass Movement**

- Overcropping occurs when soils in areas that are not suited to tillage are exposed to wind and rain.
- In the 1930s, when the region known as the Dust Bowl in the USA was changed from being a grassland region to a tillage region, its soil was exposed to strong winds that literally blew the soil away.
- The **Sahel in North Africa**, south of the Sahara, became vulnerable to erosion when overgrazing by cattle herders, overcropping by tillage farmers and drought conditions left the soil exposed to dry winds from the Sahara. This led to a cycle of drought, famine, disease and loss of life.

OR

RIVER PROCESSES

■ **The Impact of Hydroelectric Dams**

- Hydroelectric dams are built across river valleys to dam up water for the purpose of electricity generation: but this process interferes with the natural processes of river action. Rivers are forced to deposit sediment in the reservoir lake behind the dam. Eventually the dam becomes filled with sediment that should otherwise have been washed to the middle and lower stages, providing gravel, sand and alluvium supplies.
- Farmland behind the dam is likely to be flooded, and farmers' homes are submerged by rising reservoir waters.
- If a delta did exist at the river estuary before dam construction, it will now be prone to erosion due to loss of alluvial soils in the reservoir lake upstream.

- Natural vegetation in regions behind dams is lost as the reservoir waters rise.
- Villages with all their historical character may be flooded and lost under the reservoir water. This has occurred at the Three Gorges dam project in China.

■ The Impact of Canalisation

The diversion of fresh water from rivers for the purpose of irrigation can impact on natural processes in the following ways.

Increased salt content

- Many minerals are dissolved in river water; we call them salts. When salt content increases, either in the soil or in water, it can have serious effects.
- Some seas may be inland seas and are just huge, freshwater lakes.
- These inland seas, such as the Aral Sea, need a constant supply of fresh water to counteract evaporation.
- Evaporation creates a build-up of salts.
- This changes the seas' ecosystems, leading to fish species and natural vegetation being lost.

Improved agricultural output

With increased water supplies through irrigation from reservoirs, local regions that would otherwise be desert may become major farming regions, such as central and southern California.

Loss of fresh water

- Rivers that once flowed strongly to the sea may now be just a trickle.
- This causes tidal waters to reach farther up their estuaries than before dam construction. Ecosystems may be wiped out in their estuaries.

■ Flood Control Measures

Levee construction

- Levees are high banks of clay and stone, built parallel to a river's channel to contain floodwaters.
- Floodwaters no longer spread across a river's flood plain during times of heavy rain. This process naturally provided a flood plain with minerals needed by grasses to grow naturally and healthily. Levees deny a flood plain's natural mineral supply, which must now be provided by farmers themselves.
- Wildlife that once lived in marsh or wetland sections of the flood plains must find other nesting places.
- The bursting of levees can lead to severe loss of life.

OR

COASTAL PROCESSES

■ **The Impact of Recreation**

The building of groynes and breakwaters

- Demand for leisure and for coastal location of hotels has led to the creation of beaches by building groynes to trap sand along many seashores, such as along the coast of Mediterranean Spain.
- Longshore drift is interrupted and coastal currents deposit sand, leading to the creation of new beaches.
- However, the addition of sand in one region leads to the loss of sand in another, which may lead to the erosion of existing natural beaches.

Water quality

- Naturally clean water may become polluted as recreation centres increase along a seashore. Water bodies can accept a certain level of pollutants and remain clean, since natural processes are able to break down such pollutants as human waste. However, when pollutants are increased then seas become polluted, leading to poor water quality and contaminated beaches.

Coastal construction

- The construction of hotels and holiday homes along sand dunes and sand-spit environments leads to a change in natural ecosystems and damage to the coastal environment. Increased human traffic damages coastal grasses, leading to sand-dune erosion.

■ **Coastal Defence Work**

Coastal defences

- An increase in coastal storms and surges as a consequence of global warming has led to great damage in some coastal areas, such as at Rosslare Strand in Co. Wexford.
- Groynes and breakwaters are used to reduce such storm damage by breaking the force and size of waves as they approach a shore.

Sand-dune management

- Wave energy is released by waves as they crash and run up a beach. It is a natural process that keeps coastal regions in balance – so beaches should be preserved in their natural form in order to maintain this balance.
- Removal of sand by people, traffic or any measure that interferes with this balance must be eliminated or reduced to protect coastal environments.

Natural wildlife protection

Mudflats, sandflats, coastal marshes or other environments that support wildlife should be protected as part of nature's heritage. They add to the coastal attractions for bird watchers and tourism.

CORE TOPIC 12
MAPS AND PHOTOGRAPHS

DRAWING SKETCH MAPS FROM ORDNANCE SURVEY MAPS

Case Study 1: *Inishowen Region*

Examine the Inishowen extract on page 56. Then draw a sketch map (*not* a tracing), and on it mark and name its **physical regions**.

TIPS

1. Always draw a frame similar in shape to that of the map.
2. **Never** trace a map. A sketch map must be drawn freehand.
3. **'Mark'** and **'name'** (or 'label') are different directions, and marks will be awarded for each separately.
4. Never draw a very large sketch map, as it is more difficult to draw and it takes up too much time.
5. Practise different types of sketch map, and time yourself.
6. Use a soft lead pencil.
7. Use colour **only** if you have enough time.

To identify individual physical regions on a map it is often helpful if you squint your eyes; the separate regions may then become clearer.

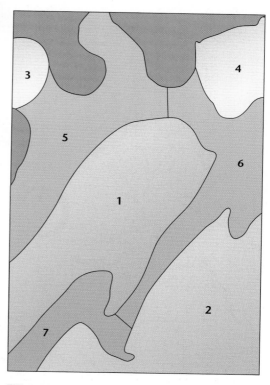

▲ Sketch map of Inishowen map extract

Key: physical regions

1. Raghtin More upland ridge
2. Bulbin uplands   }High uplands
3. Dunaff uplands
4. Binnion uplands   }Low uplands
5. Dunaff coastal plain
6. Basin of the Ballyhallan–Clonmany Rivers
7. Basin of the Owenerk River   }Lowlands

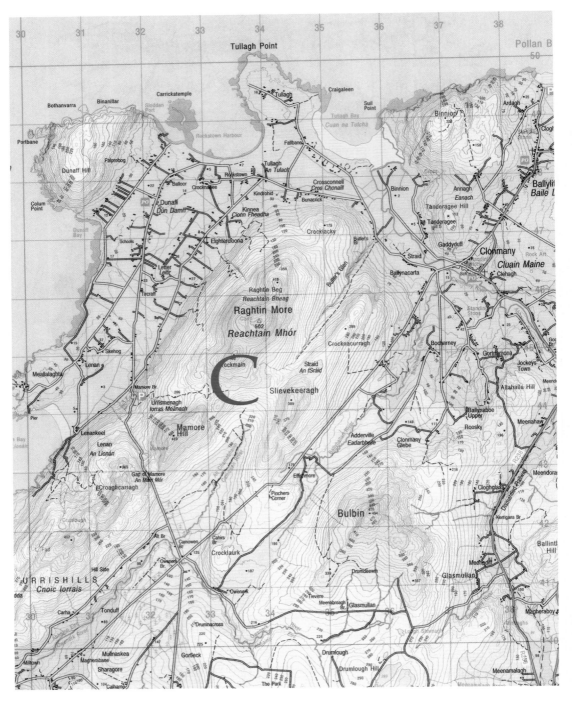

▲ Inishowen map extract

Case Study 2: *Munster Ridge and Valley Region*

Study the Lismore–Cappoquin map extract on page 58. Then on a sketch map:

(i)   divide the area of the map into its physical regions,

(ii) mark and name three urban regions, and

(iii) mark and name the main rivers.

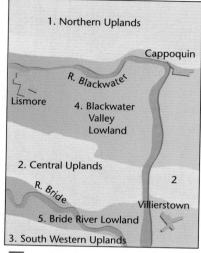

▲ Sketch map of Lismore–Cappoquin

Key: physical regions

1. Northern uplands
2. Central uplands
3. South western uplands
4. Blackwater river valley
5. Bride river valley

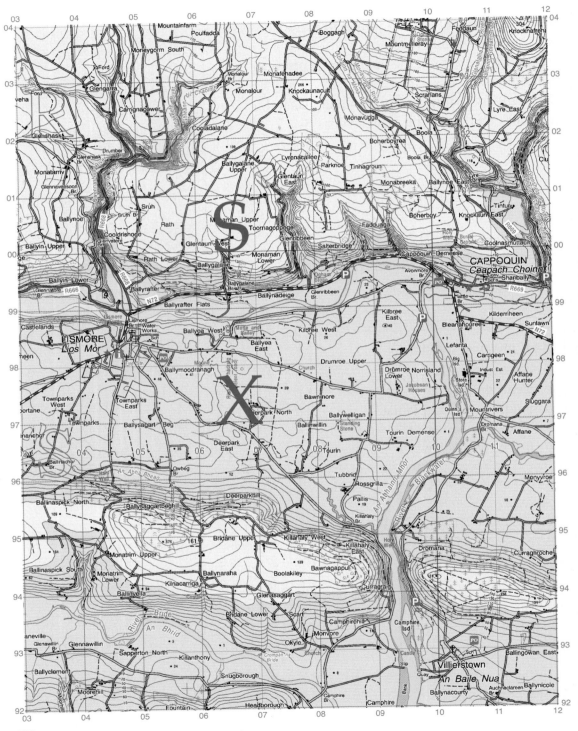

▲ Lismore–Cappoquin map extract

Case Study 3: *Sligo Coastal Region*

Study the map of the Sligo Coastal Region on page 60. Then on a sketch map mark and name:

(i)   the physical regions,

(ii) Sligo urban region, and

(iii) a coastal resort region.

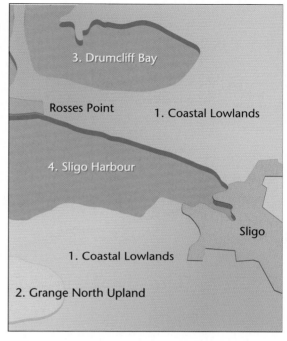

3. Drumcliff Bay

Rosses Point

1. Coastal Lowlands

4. Sligo Harbour

Sligo

1. Coastal Lowlands

2. Grange North Upland

▲ Sketch map of Sligo coastal region

Key: physical regions

1. Coastal lowlands
2. Grange North upland
3. Drumcliff Bay
4. Sligo Harbour
5. Sligo Town urban region
6. Rosses Point resort region

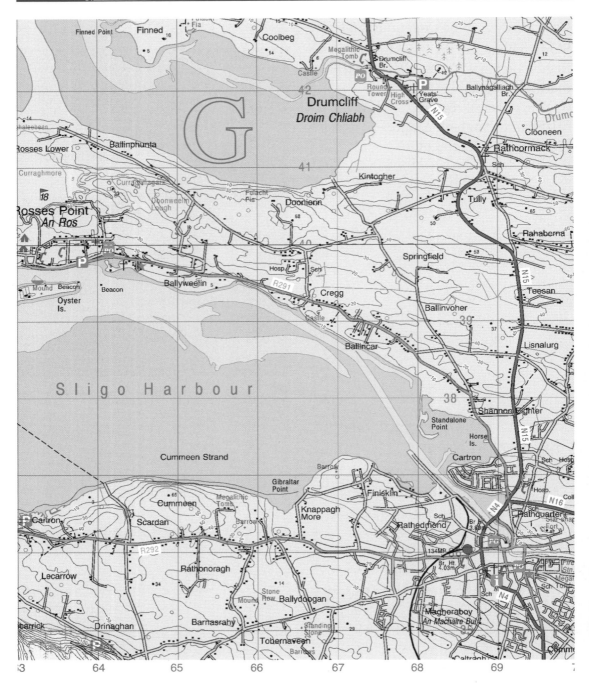

▲ Sligo coastal region map extract

Case Study 4: *The Swords–Malahide–Donabate Region*

Study the map of the Swords–Malahide–Donabate region on page 62 Then on a sketch map mark and name:

(i)   the main urban regions,
(ii) a recreational region,
(iii) two nature reserves,
(iv) an agricultural region, and
(v) a woodland region.

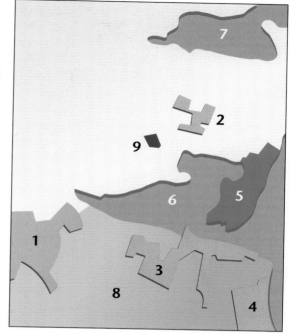

▲ Sketch map of the Swords–Malahide–Donabate region

Key: regions

1. Swords urban region
2. Donabate urban region
3. Malahide urban region
4. Portmarnock urban region
5. Golf links
6. Malahide Bay Nature Reserve
7. Rogerstown Estuary Nature Reserve
8. Agricultural region
9. Newbridge Demesne Woodland

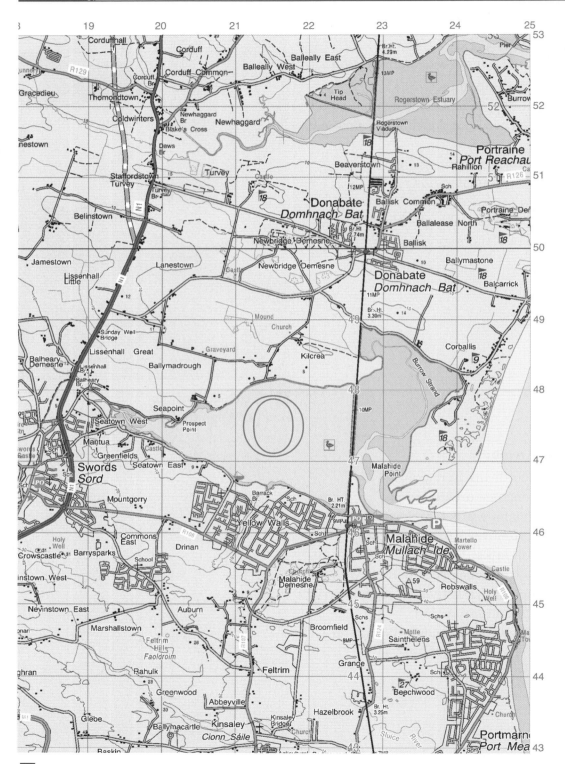

▲ Swords–Malahide–Donabate map extract

Case Study 5: *The Nore River Valley*

Study the Nore map extract on page 64. Then on a sketch map (not a tracing) of this region, mark and label the following:

(i)   the upland regions,
(ii) the highest point,
(iii) the river Nore,
(iv) one regional road,
(v)   three third-class roads, and
(vi) one urban region.

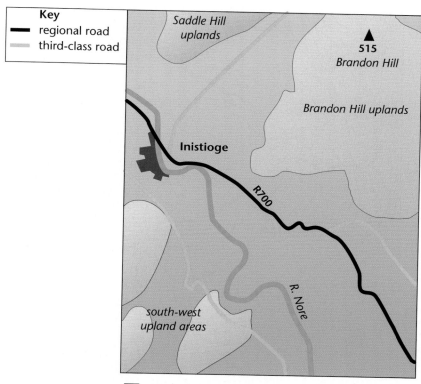

Sketch map of Nore River valley

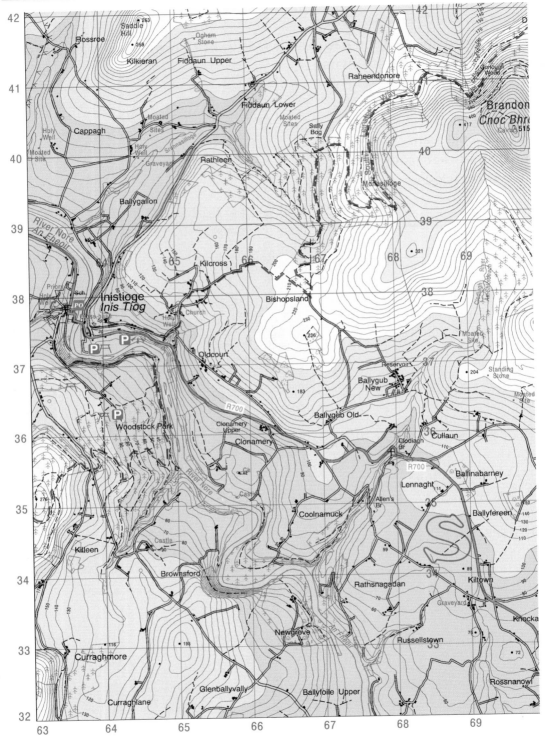

▲ Nore River valley map extract

Case Study 6: *Donegal Region*

Study the photograph of the Donegal region on page 66. Then on a sketch map mark and name the following:

(i)   an urban region,
(ii) two residential regions,
(iii) the central business-district region,
(iv) two industrial regions, and
(v)   a green belt region.

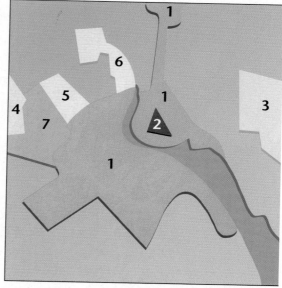

▲ Sketch map of the Donegal region

Key: regions

1. Donegal urban region
2. Central business district
3. Residential region
4. Residential region
5. Industrial region
6. Industrial region
7. Green belt region

Aerial photograph of the Donegal region

DRAWING SKETCH MAPS FROM AERIAL PHOTOGRAPHS
Case Study 7: *Donegal Town*

Study the photograph of Donegal town on page 68. Draw a sketch map (*do not trace*) of the area shown, and on it mark and label:

(a) the street pattern of the town
(b) five regions of different land use in the town
(c) the river.

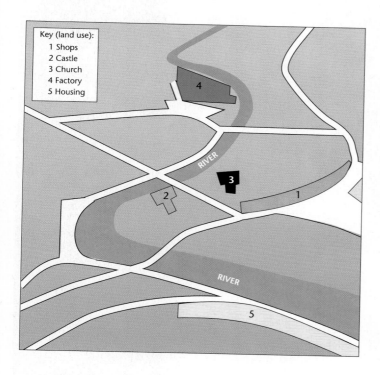

Key (land use):
1 Shops
2 Castle
3 Church
4 Factory
5 Housing

Key (land use)

1. Shops
2. Castle
3. Church
4. Factory
5. Housing

◀ Sketch map of Donegal town

TIPS

1. Always draw a frame similar in shape to that of the map.
2. Never trace a map. A sketch map must be drawn freehand.
3. Keep the sketch size to less than half of a sheet of foolscap paper.
4. Show and name only the features that you are specifically asked for.
5. Always outline your sketch with a soft pencil. This allows you to correct errors that may arise.
6. Outline land-use zones with a heavy boundary line to limit the area.
7. Always mark and label each land-use area.
8. Use colour only if you have enough time.

▲ Aerial photograph of Donegal town

Case Study 8: *Ballybunnion: A Coastal Town*

Examine the aerial photograph of Ballybunion on page 70. Draw a sketch map based on the photograph (you may *not* use tracing paper), and on it mark and label:

(a) two areas where coastal erosion is evident

(b) two areas where coastal deposition is evident [**only areas asked for here; if in doubt add in feature name also**]

(c) three regions of different land use that are related to the coastal location of this settlement [**in this case land uses must be directly related to a seaside area**].

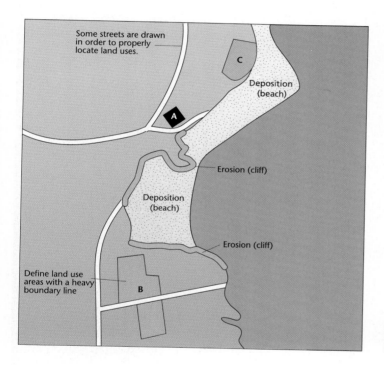

Some streets are drawn in order to properly locate land uses.

C

Deposition (beach)

A

Erosion (cliff)

Deposition (beach)

Erosion (cliff)

Define land use areas with a heavy boundary line

B

Key (land use)

A. Hotel

B. Mobile-home site

C. Pitch-and-putt course

These land uses are regularly found in seaside resorts.

◀ Sketch map of Ballybunion

TIPS

1. Use a key or use labelled arrows to avoid overcrowding.

2. Limit sketch size to half a foolscap page to save time.

3. Always mark and label features to be identified.

4. Carefully examine the questions asked, and include only what is asked of you. If in doubt, add one extra example.

▲ Aerial photograph of Ballybunion

CORE UNIT 2: REGIONAL GEOGRAPHY

CORE TOPIC 13
REGIONS

■ **What is a Region?**

A region is an area of the Earth's surface that has human and/or physical characteristics that give it an identity and that make it different from all the areas around it.

The different types of region include:

Climatic regions
Physical regions
Administrative regions
Cultural regions
Socio-economic regions
Urban regions.

CLIMATIC REGIONS

- Climatic regions are areas that have their own climate, distinct from those regions surrounding them.
- Some climate regions are huge, such as the Sahara Desert, while others are small, such as individual island climates.

■ **Cool Temperate Oceanic Climate**

Western Europe – which includes Ireland, Britain, Scandinavia, Denmark, Netherlands and the west coast of France, Spain and Portugal – has a Cool Temperate Oceanic Climate.

Temperature

Summers are warm: 15°C to 17°C.

Winters are cool. January temperatures average 4°C to 5°C.

Precipitation (rainfall)

- Rain falls throughout the year, but most falls in winter.
- Relief rain falls in mountain regions.
- Western areas, such as the west of Ireland, receive more rain than eastern areas such as Dublin.
- Cyclonic rainfall occurs because of depressions that travel across the ocean between 30°N and 60°N.

Winds

The South-West Anti-Trades are the prevailing winds of cool temperate regions in the northern hemisphere. The North-West Anti-Trades are the prevailing winds of cool temperate regions in the southern hemisphere.

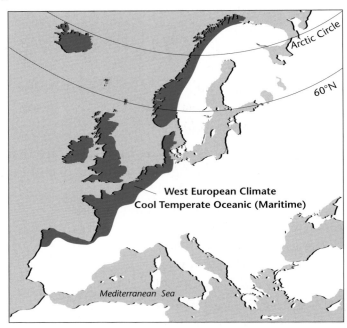

▲ Western European climate

PHYSICAL REGIONS

- ■ **Karst landscapes – The Burren in Co. Clare**
- Karst landscapes are those where large expanses of weathered limestone rock are exposed at the surface.
- The Burren is an upland region that was uplifted when the African and Eurasian plates collided during the Armorican period, about 300 million years ago.
- It has large areas of exposed limestone due to erosion, because of overcropping and overgrazing by Ireland's earliest farmers.

Karst landscape surfaces include surface features such as sinkholes, dry valleys and limestone pavement with its grikes and clints.

Other karst landscapes in Ireland include the **Dartry–Cuilcagh uplands** in counties Fermanagh and Cavan.

■ Munster Ridge and Valley Province

- This area is a distinctive region. Its parallel sandstone ridges and limestone valleys run east–west across southern Munster.
- These ridges and valleys were formed by the same earth movements that uplifted the Burren in Co. Clare.
- The northern boundary of this region is the **Armorican Thrust Front**. This boundary separates the severely folded rock to the south from the gently folded bedrock to the north.
- The coastal edges of these east–west valleys dipped into the sea when their support was lost as the American plate moved away from Europe. These inlets, such as Dingle Bay, are now called **rias**.

■ North European Plain

- This is a lowland region that stretches from northern France to Bulgaria and the Black Sea. The region was levelled by sediments that were eroded from surrounded mountain and upland regions and deposited by wind and rivers.
- Parts of this lowland were covered by rising sea levels after the ending of the Ice Age some 10,000 years ago. So Ireland and Britain became islands and were cut off from mainland Europe.

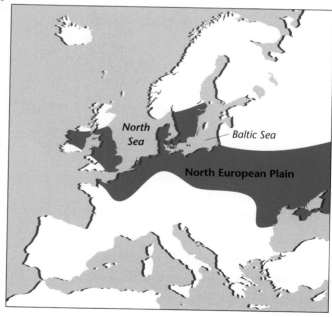

▲ The North European Plain

ADMINISTRATIVE REGIONS

Administrative regions need to be big enough to provide services efficiently. They also need to be small enough to work effectively and reflect community interests.

Counties, city councils and borough councils in Ireland ▶

- Systems of Government Administration

Single-tier system

In Ireland local government, such as city councils and county councils, has direct access to the central government in Dublin

- Administrative Units in Ireland

The counties of Ireland

- The county was central to a system of administration introduced by the Normans as they advanced across the country by military conquest. Counties are defined by boundaries such as rivers and mountain ridges.

Urban-based administrative units

- *City Councils*

Dublin, Cork, Limerick, Waterford and Galway.

- *Borough Councils*

Clonmel, Wexford, Kilkenny, Sligo and Drogheda.

- *Town Councils*

There are 75 town councils. These were established in the nineteenth century.

Regional administrative units

Health boards, industrial development authorities and other organisations all have different regions under their control. This leads to poor planning and inefficiency in Ireland.

Multiple-tier system

In France local governments, called departments, have indirect access to central government in Paris, through government officials and regional governments.

Local government in Ireland

There are over 100 local authorities in Ireland.

The role of Local Authorities

- dealing with local issues at a local level
- providing essential services such as sewage, water and housing for those who otherwise could not afford a home. Other services include education, refuse collection and some recreation and amenities
- drawing up strategic development plans for future needs and preservation of the natural and built environment
- maintaining transport routes.

Local planning operates under three headings:

Subsidiary – this involves decision making at local level to encourage self-reliance.

Appropriateness – services and administration to the highest standards provided in accordance with local needs by local people, so that the state can function efficiently.

Partnership – encourages local people to take part in government.

■ Local Government in France

The 'Départements' of France

France is divided into **92 'départements'** which are all roughly the same size.

- After World War II clusters of départements were organised into regions.
- In 1982 a new law gave each region new status to overcome the powerful influence of France's primary city, Paris.

- French regions now have responsibility for:
  - economic and cultural activities, such as job creation, tourism and heritage in their own regions;
  - effective planning and the coordination of new initiatives proposed by the local government and financed still by the state;
  - people voting in direct elections to new regional assemblies or 'councils'.
- Some regions have become quite powerful, such as Lyon–St Etienne–Grenoble, and help to counterbalance the dominance of Paris.

CULTURAL REGIONS

Language and religion are two factors that help to define certain cultural regions.

■ Language Regions

- Language plays a major part in defining cultural identity.
- It is passed from generation to generation and even within immigrant communities it creates cultural regions within cities: for example, Little Italy in New York, or Chinatown in San Francisco.

Gaeltacht regions in Ireland

- 1.6 million Irish people claim to have some ability to speak Irish.
- It is only in Gaeltacht regions that it is used in everyday life.
- In 1925 Gaeltacht regions were divided into two categories:
  - **Fior Gaeltacht** regions, where 80 per cent or more of the population speak Irish
  - **Breac Gaeltacht** regions, where 25–79 per cent of the population speak Irish.
- Gaeltacht regions have since reduced in size and number and are confined to scattered regions along the west and south coasts. They have a total population of 86,500 people.
- They retain a strong bond with the Irish people.

Language regions in Belgium

Belgium has three language regions:
- Northern Belgium speaks Flemish, a language related to Dutch
- Southern Belgium speaks French, but also includes a small German-speaking community
- Brussels is the capital city, where Flemish and French have equal status.

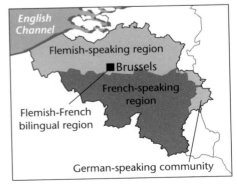

▲ The language regions of Belgium

Flemish-speaking region

- Northern Belgium, called **Flanders**, was a region of small farming communities in the nineteenth and early twentieth centuries. Today it is a rich industrial region. Yet it feels under threat from the more dominant international French language.

French-speaking region

- Southern Belgium is called **Wallonia** and was once a rich coal-mining area. It is now struggling to compete for new industries and has a high unemployment rate.

Tensions between these communities led to fundamental government reforms, and Belgium now has a federal-style government which recognises the three regions above based on language.

■ Religious Regions

Some regions may be defined by the religious beliefs or characteristics of their population, or by religious conflict between different religious groups.

Northern Ireland – A region of religious conflict

- The establishment of the Irish Free State in 1921 left six counties of Ireland under British rule.
- There was a majority of Protestants living in four of the counties.
- Two additional counties were included to make the six counties economically viable as a separate political unit.
- Since partition, conflict has continued between the Catholic minority and the Protestant majority.
- Religious communities are also divided within urban regions such as Derry and Belfast, creating Catholic-only and Protestant-only ghettos.
- Examples include the Catholic-majority Falls Road community and the Protestant-majority Shankill community in Belfast.

The Islamic World

- The Islamic World includes all North Africa and South West Asia.
- It mostly coincides with a vast, hot-desert climatic region that includes the Sahara and Arabian deserts.
- Powerful Arab armies and Arab traders converted the populations of these desert regions to Islam.
- **Mosques**, with their characteristic towers called **minarets**, are characteristic of Islamic landscapes. Islamic invaders into Europe were called **Moors**; they conquered Spain, but their northward advance was stopped at the Battle of Tours.

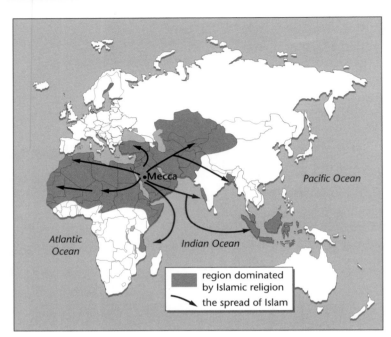

◀ The Islamic world

region dominated
by Islamic religion

the spread of Islam

SOCIO-ECONOMIC REGIONS

■ EU Regional Funding

In 1998 the EU's Common Regional Policy was changed to deal with increasing inequality between richer and poorer regions. Increased funding was given to three categories of region.

Objective 1

- These regions were defined as having a GDP (Gross Domestic Product) per person of less than 75 per cent of the EU average.
- These are EU regions with the most problems and they need the most support.
- They are generally large areas, such as Northern Sweden, the Mezzogiorno in Italy and BMW in Ireland.

Objective 2

- Funding is given to help old, urban-industrial regions cope with the loss of their traditional industries such as coal-mining and iron and steel industries, for example in the Sambre–Meuse Valley in Belgium.
- Funding was also given to attract industries to less-developed rural areas, such as the Massif Central in France.

Objective 3

- These funds apply throughout the EU. Their purpose is to help marginal groups of people, such as ethnic minorities, the handicapped or unemployed young people, to become better integrated into society and to find jobs.

■ 1. Core Regions

Core regions are wealthy regions. They may occur within a country, such as Ireland, or within an economic region such as the European Union.

A. The Dublin region

- Dublin is Ireland's core region. It is the country's capital, its largest centre of population and services and has well-developed transport and communications systems.
- Over one-third of all full-time jobs in foreign-owned manufacturing and financial services are in Dublin.
- The city is Ireland's major port, and the location for financial and commercial company headquarters.

B. The Core of the European Union

- Because the countries of Western Europe are relatively small, a number of national cores and growth centres have combined to create an international core. It is called the **'European Dogleg'** or **'Hot Banana'**.
- The core also includes the 'Four Motors' or industrial regions of the European Union: Stuttgart, Lyon, Barcelona and Milan.

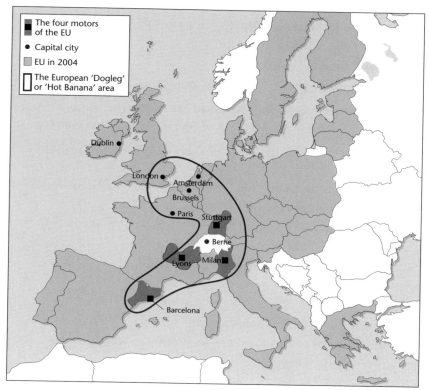

▲ Europe, showing the 'dogleg'

■ **2. Peripheral Regions**
- Peripheral regions are generally located at the edge of the EU.
- They include the BMW region in Ireland, north-west Scotland, Spain, the Mezzogiorno in southern Italy, and Greece.
- They suffer either from rural underdevelopment or industrial underdevelopment, or both.
- Other problem regions include those in industrial decline, such as the Sambre–Meuse Valley coal-mining region in Belgium.

(i) The Border–Midlands West (BMW) – Ireland's problem region
- The disposable income of the Border, Midland and Western region is 9 per cent lower than the national average.
- It has Objective 1 status in the EU for the period 2000–2006 and so is able to benefit from structural funding.
- To qualify, it must have a GDP (Gross Domestic Product) per person of less than 75 per cent of the EU average.
- Much of its western edge is mountainous, with blanket bogs, and it is liable to flooding in the midlands.
- Primary activities are dominant (see BMW region, pages 84 to 88).

(ii) The Mezzogiorno in southern Italy (see pages 91 to 95).

■ **3. Regions in Industrial Decline**

(i) Example: The Sambre–Meuse region in Belgium
- Since 1750 and up until the 1950s, traditional industries such as coal mining and iron and steel factories were located on or close to coalfield areas.
- This led to large-scale, heavy industrial regions such as the Sambre–Meuse region in Belgium.
- The Sambre–Meuse is an Objective 2 region and its coalfields stretch for 150 kilometres along the Sambre–Meuse valley.
- Its industries include the heavy industries such as iron and steel, engineering and chemicals. However, owing to:
  - new sources of energy such as oil and natural gas
  - cheaper imports
  - new materials such as plastics to replace metals
  - new technologies and newer, more efficient factories in coastal locations, the competitiveness of old industries declined and so did the region that depended on them. This process is called 'deindustrialisation'.

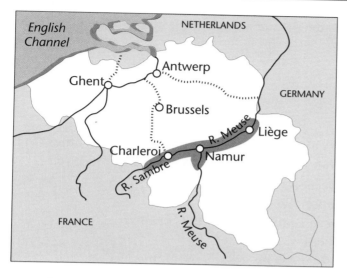

◀ The Sambre–Meuse Valley region

Improvements as a result of Structural Funds
- New motorways that link the Sambre–Meuse to neighbouring urban industrial regions.
- New industrial estates along the new motorways and near the large cities of the region.
- Improvements at Charleroi airport.
- Cleaning up and planting of conifers on old slag heaps to improve visual landscape.
- New industries, such as Caterpillar at Charleroi and Ford at La Louvière have been attracted to the region.

(ii) Example: The Greater Cork region
- In 1973 Cork was Ireland's leading port-related industrial region.
- Its industries included Irish Steel, Verholme Dockyard, Whitegate Oil Refinery and Dunlop and Ford car-assembly plant. An international recession led to closures at Ford, Dunlop, Verholme and Irish Steel, with a loss in excess of 3,000 jobs.
- Since the 1990s government and regional planning have attracted new industries. These include chemical and pharmaceutical companies, such as Pfizer, around its large harbour.
- IT companies have set up in the many industrial estates around the edge of the city.
- Increased investments in the following have changed the image and wealth-creating ability of the Cork region:
  - educational centres such as University College Cork, Cork Institute of Technology

> – port facilities such as the deepwater container and passenger ferry facilities at Ringaskiddy
> – improved roads
> – urban renewal.

URBAN REGIONS

An urban region includes a town or city and the surrounding area that is linked to it by activities such as shopping, journeys to work and supplying farm produce.
Urban regions are important to us because:
- 50 per cent of the world's population lives in cities or towns
- 80 per cent of the people in Western Europe live in cities or towns
- 60 per cent of Irish people live in cities or towns
- People's lives are increasingly influenced by urban environments.

■ **Ireland's Urban Regions**

- Ireland is one of the least urbanised countries in Western Europe.
- The eastern and southern parts of the country are the most urbanised.
- Our eastern and southern coastal towns and cities were established by the Vikings, and the inland towns and cities were established by the Normans.
- Dublin is a primate city because it has about ten times the population of the next largest city, Cork City.
- Its urban environment extends as far as Athlone to its west, Dundalk to its north and Arklow to its south.
- New motorways have increased commuter traffic to work on a daily basis in each of these directions.

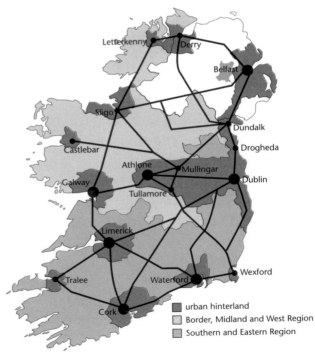

▲ Ireland's urban regions

■ European City Regions

About 80 per cent of the population of Western Europe lives in towns or cities. There are three major zones of urban settlement in Western Europe. These are:

- **Manchester–Milan axis** – based on a historic trading route that joins the North Sea and the Rhine waterway to the Mediterranean.
- **Paris–Berlin axis** – follows a rich agricultural lowland region called the Hellweg, which runs between the North Sea and the upland regions to the south.
- **Along the coastline of Western Europe** – at the lowest bridging point of rivers on deep coastal estuaries. Major ports include Le Havre, Antwerp, Rotterdam and Hamburg.
- Two multi-centred urban core regions have developed where the Manchester–Milan axis intersects with the Coastal axis.
- These are the Randstad conurbation in the Netherlands and the Rhine–Ruhr conurbation in Germany.

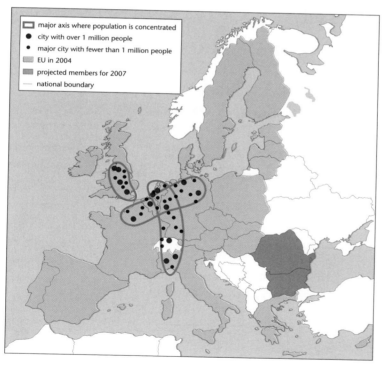

◀ Europe showing principal city regions in the core

The Paris city region

- Paris is located in the centre of the Paris Basin at a focus of routes on the River Seine. It began on an island in the Seine, the Île de Cité.
- It is a confluence town, where the rivers Marne and Seine meet.

- It is a major nodal centre, the most important industrial centre, and the capital city of France for over a thousand years.
- The Paris Basin that serves the city is the richest farming region in France.

Recent planning developments in Paris

- **Suburban growth centres** – Eight suburban centres where homes and places of work were developed to reduce commuting distance and traffic congestion and create employment. La Défense was the first and largest of these suburban nodes to revitalise Paris.
- **New towns** – Five new towns were built along two axes to the north and south of the River Seine.
- **New growth centres** – These are based on existing cities within the Paris Basin, such as Rouen and Orléans, and were established to restrict the dominance of Paris.

CORE TOPIC 14

CONTRASTING REGIONS IN IRELAND AND EUROPE

THE BORDER, MIDLAND AND WEST REGION

■ **Factors that created the Environment of the West of Ireland**

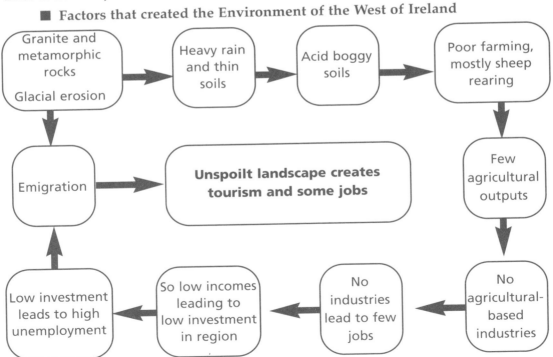

■ Relief and Soils

- The BMW region is a broad, central lowland enclosed by a broken arc of upland and mountains.
- Much of the western coastline was submerged by the sea, so it has many bays and inlets.
- Much of the western edge of the BMW region has rugged uplands, where glacially eroded mountains rise higher than 300 metres.
- Its mountains and uplands include the Connemara, Mweelrea, Nephin Ox, Cuilcagh Uplands.
- The western part of the central undulating lowland is composed of the river flood plains of the Rivers Clare, Moy, Shannon and Suck. Many lakes also exist. They include Loughs Conn, Mask, Corrib and Ree. Much of the lowland soil is poorly drained.
- Glacial deposits disturbed natural drainage by glacial action and shallow depressions were created that filled with water to form lakes in which deep, raised bogs formed, close to the Shannon river.
- The soils of eastern Galway are shallow, with carboniferous limestone close to or exposed at the surface.

■ Rock Surface

- Much of the western edge of the BMW region has either granite or metamorphic rock exposed at the surface.
- Both these rock types produce poor soils.
- The poor soils and the heavy rain create a cycle of poor soil, few farming activities and low output. These in turn lead to low incomes.

■ Climate

- The climate is Cool Temperate Oceanic. It is relatively mild throughout the winter months, with January temperatures averaging about 4°C to 5°C.
- The North Atlantic Drift, which flows from the Gulf of Mexico towards Ireland, has a moderating influence that keeps winter temperatures mild and keeps the seas ice-free.
- Blowing over this warm water surface are the South-Westerly Anti-Trades, which bring warm air to coastal regions. Summers are warm, with average temperatures rarely above 15°C.
- The climate is mild, wet throughout the year and very windy. This is directly linked to the prevailing south-westerly winds and frontal depressions which are forced to rise over the mountainous western

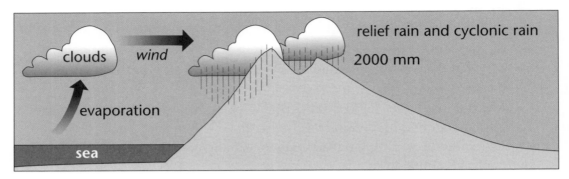

▲ Rainfall in the west of Ireland

coastline. These create relief rain over the mountains. Precipitation can be higher than 1,500–2,000 mm annually, with more than 250 days of rain in the year, mostly falling in winter.

■ **Primary Activities**

Farming

Because the Industrial Revolution had little impact on Ireland, agriculture is the most important single industry in the economy of the Republic of Ireland.

- Traditional farming in the west of Ireland provides only low income to most farmers. Average farm income is only 50 per cent of that of the eastern region, and only 14 per cent of farms can be considered viable, full-time units.
- Difficult environmental conditions, namely high rainfall, peat and waterlogged soils, thin stony soils and mountainous terrain, limit how productive the land can be for agriculture.
- 63 per cent of Irish farms are located in the BMW region. The average size of farms is small at 18.4 ha, mechanisation is low and a high proportion of farmers are older. Tillage is not suitable for most of the region. Grazing of beef cattle and sheep is dominant.

Forestry

- Even though more than 50 per cent of Ireland's land is more suited to forestry than agriculture, only **less than 10 per cent** is forested.
- Ireland's long growing season, well-distributed rainfall and mild temperature gives a growth rate for trees more than three times higher than the continent of Europe.
- The EU has promoted forestry as a more profitable activity than farming in marginal regions such as the Border, Midlands and West.
- The state is hoping to cover up to 15 per cent of Ireland with forests.

Fishing

- The continental shelf stretches for approximately 480 kilometres off the west coast of Ireland. It is **a shallow sea** region where sunlight can penetrate to the ocean floor. This encourages the growth of **plankton, a microscopic food for fish**. The warm waters of the Gulf Stream attract a large variety of fish, such as herring and cod.
- Our sheltered, ice-free waters can be fished year-round. The growing demand for fish and fish products has created jobs and wealth for many in our fishing ports of the West. This has occurred despite problems of **overfishing** and **reduced fish quotas**. Our most important fishing ports are on the west coast. They include **Killybegs, Castletownbere**, Rossaveal and Dingle.
- The value of fish landings at the important fishing ports on the west and north-west coasts is approximately €45 million annually.
- Aquaculture is a major growth industry. The very irregular coastline and sheltered bays and estuaries of the west, with pollution-free waters, form an ideal environment for this industry.

■ Secondary Activities

- Ireland's industrial revolution began in the early 1960s, when a large number and range of industries set up in newly established industrial estates that catered for industry only.
- Ireland's peripheral location as an island on the western edge of Europe, together with its lack of quality coal and discovered metals, caused this late start; the process had begun as early as the 1750s in Britain.
- **Membership of the EU in 1973, improved trucks, roads, ferries and airports have helped boost competitiveness.** Branch plants of foreign multinationals located their factories in rural areas because land and labour were cheap.
- Government grants and tax incentives also encouraged this trend.
- During a recession in the 1980s many uncompetitive branch plants closed or reduced their workforce.
- **High-tech multinationals** were not attracted to rural regions. However, some urban regions became attractive, such as **Galway City** because of its university status.
- Dependence upon foreign companies is high, with 51 per cent of the workforce employed in their factories. This is a concern for many communities that depend heavily on a single foreign company for employment in their area.

This region has an **Objective 1 status**. That, and government support, are vital to improve the region's road, rail, air, telecommunications and general urban facilities to make it more attractive for new, modern industries.

■ Tertiary Activities

The majority of people in a developed economy are employed in the tertiary sector. These services are available in urban regions.

- The **BMW region is mainly rural** and so the number of people employed in the service sector is lower than the national average. This is because the rural society associated with agriculture is less attractive for companies than urban regions that have a manufacturing tradition.
- Apart from Galway, most towns in the BMW region do not provide a good range of high-quality services. Many people from the region commute to urban areas, such as Galway, Dublin and Limerick, for work.
- The **decentralisation** of government departments to regional centres, such as the Department of Education to Athlone, helps to reduce the imbalance of service employment.
- The BMW region has many advantages for **tourism**, such as scenic areas like Connemara, Clew Bay, the Shannon waterway. Despite this the tourism industry remains underdeveloped.
- The BMW region has 52 per cent of the bed capacity, but generates less than 40 per cent of the country's tourist revenue.
  - **Tourism is seasonal**, with July and August the most important months.
  - Many in the catering and hotel trade become **unemployed in the off-season.**

■ Human Processes

- The population of the BMW region has been in decline since famine times.
- Migration has led to **more older people** than younger people in the population, resulting in a 'brain drain'.
- This is an additional factor in the region's being less attractive for the location of industries.
- More than 18 per thousand of the population in Connacht are aged over 60 years. Death rates are 10.8 per thousand.
- Although it covers 60 per cent of the country, it has only one-third of the population.
- Over **66 per cent of the people live in rural areas.**
- Most of the Gaeltacht areas are located in this region.
- The boundaries of these Gaeltacht regions have reduced, as have their populations.
- The region is the heart of the Irish culture.

THE SOUTHERN AND EASTERN REGION

■ Relief and Soils

- Most of the land in this region is undulating lowland. Brown soils formed from limestone glacial drift cover much of the region, and these include some of the most fertile soils in the country.
- Some mountain chains are found in this area. They include the Wicklow, Blackstairs, Comeragh and Knockmealdown mountains.
- Rivers are wide and deep and drainage of the region is much better than in the BMW region. The rivers flow into the Irish and Celtic seas. They provide natural, sheltered inlets at their estuaries, where their lowest bridging points have given rise to our largest cities such as Dublin, Cork and Waterford.

■ Climate

- Again the climate is Cool Temperate Oceanic, but less severe than in the West. The lower level of the land and the rain-shadow effect of the western mountains result in much lower rainfall levels: less than 1000 mm in places.
- Rainfall is better distributed throughout the year than in the West. The south-westerly winds are milder and less severe owing to the presence of trees, which are absent along the west coast.
- Winter temperatures are colder because of the effect of increased distance from the milder Atlantic Ocean. However, summer temperatures and average amounts of sunshine hours are greater than in the West.

■ Primary Activities

- Because soils are deeper and richer, farming is more productive than in the West and Midlands. Farms are highly mechanised, larger and have a higher percentage of younger, more energetic and more innovative people.
- Farming such as cereal growing and dairying are intensive and specialised, giving higher incomes for farmers. The average farm income is 40 per cent above the national average. The greater diversity of natural environments allows for a wide range of agricultural production.
- Tillage is common in the east and south-east, where soils are deep and well drained and sunshine levels are highest.
- Forestry is confined to upland slopes, where soils are thin and more marginal.
- Only Howth and Dunmore East can be classed as major fishing ports.
- Fishing vessels have the disadvantage of having to travel longer journeys to find rich fishing waters than do the boats of the West.

- The presence of more polluted water from industries and towns and cities in the Irish Sea limits the possibilities for aquaculture.

■ **Secondary Activities**

- High-tech industries, such as computer-related companies, are strongly attracted to the east and south because our largest cities and towns are located there.
- Good communications systems, access to universities and educated workers form a huge drawing power. International transport links, such as airports and ferry services, and good recreational facilities are key locational factors for these growth industries.
- Almost 60 per cent of Ireland's net growth in manufacturing in the 1990s was in Dublin and its hinterland.
- Dependence on foreign companies has increased more than in the western region – so there is greater vulnerability if there is a global or US recession.
- Dublin is by far the most dominant location for manufacturing industry. Its hinterland spreads to the Midlands, from where many people commute to work on a daily basis.
- Because the greater Dublin area, with a population of 1.3 million people, is a primate city, it dominates our manufacturing output. Its numerous industrial estates cater for a wide range of industries such as pharmaceuticals, food processing and computer software.

■ **Tertiary Activities**

- In 1981 Ireland became defined as a service economy when, for the first time, more than half the working population was employed in the tertiary sector. By 2002 approximately 70 per cent of all employment was in services, and three-quarters of these jobs were in the eastern and southern region.
- In the 1990s four out of every five jobs created were in the services sector. The three most important of the international traded service industries (ITS) are:
  - computer software
  - data processing (including telesales)
  - international financial services.
- By 2002, some 56,000 jobs were available in ITS and Greater Dublin benefited most. Its well-developed communications systems have been vital in attracting data-processing operations, while the many third-level colleges and universities have attracted the computer software companies.

- The most notable new development in Dublin has been the **International Financial Services Centre**.
- Dublin is the country's capital city and the decision-making centre for many public and private enterprises.
- It is the dominant shopping centre, with a range of major educational, health and recreational facilities, and it is the hub of the country's transport system.
- Over **60 per cent of the €4.9 billion of tourist revenue** was spent in the southern and eastern region. This is directly related to air access, as 93 per cent of all scheduled flights to Ireland land in Dublin.
- The southern and eastern region has more developed transport systems than the BMW region, including:
    - most of the country's ports
    - two of the three international airports
    - the national rail and road networks which meet in Dublin.

■ Human processes

- The south and east has increased its share of the national population, owing to migration from the BMW region in search of work. Its population is now three times greater than that of the BMW.
- Its **towns are more numerous,** larger and more evenly distributed, and three-quarters of its population lives in these urban centres.
- An estimated 86 per cent of Ireland's third-level places are in the region, and 90 per cent of their graduates find jobs here.
- The population is younger, with only 14 per cent aged over 60 years.
- Birth rates are higher and death rates are lower than in Connacht.
- The region provides an **above-average prosperity** level, higher than the EU average.

THE MEZZOGIORNO IN ITALY:

A Hot Peripheral Region
■ Relief and Soils

- Southern Italy is dominated by the steep slopes of the Apennine mountains, which stretch 1,500 km along the spine of the peninsula.
- The Mezzogiorno stretches from Rome to Sicily.
- The rich, fertile, alluvial (river) and volcanic soils from weathered lava are mostly located in valley flood plains or narrow coastal plains.
- Calabria, in the toe of Italy, is mostly granite plateaus with poor, thin soils.

- The Apennines were formed from the collision and subduction of small tectonic plates. One seafloor plate sinks under the toe of Italy, causing volcanoes such as Vulcano in the Lipari Islands and Mount Etna in Sicily.
- The Tiber is the largest river and enters the sea south of Rome.

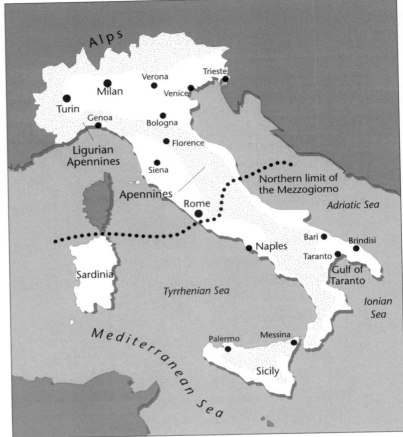

▲ Italy, showing the Mezzogiorno region

The remaining rivers are small, fast-flowing streams from the Apennines that often reach the sea through narrow gorges, especially on the western coast.
- Much of the bedrock is porous limestone that allows little surface drainage.
- The high Apennines are karst landscapes, where limestone bedrock is exposed over large regions.

■ Climate

- The Mezzogiorno has a Mediterranean climate.
- High pressure dominates in summer. Winds are hot and dry and blow outwards as north-easterly winds from the continent of Europe. These dry land winds create drought and severe evaporation throughout the summer, from June to September.
- Any summer rains fall as intense downpours accompanied by thunderstorms. These create rapid runoff and erosion, often leading to

landslides and mudslides.
- Temperatures are high, with an average of 29°C.
- Winters are mild, about 17°C and moist. Rainfall amounts vary from 500 mm to 900 mm annually.
- The lowest rainfall occurs along the Adriatic coast, because it is in the rain shadow of the Apennines.

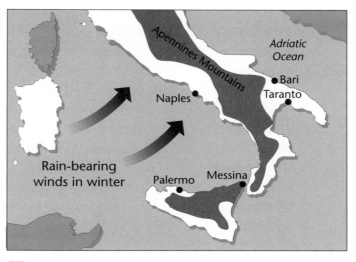

▲ Italy, showing the Apennines and rain-bearing winds

Primary Activities

- Up until the 1950s the majority of the working population was employed in farming and fishing. The people were poor and income was low. For centuries Italy was not one country, as it is now.
- The system of land ownership was called **Latifundia**, by which most of the best land was owned by absentee landlords. They leased land to peasants for grazing animals such as sheep or for growing cereals.
- The form of farming was extensive farming, where a lot of land was farmed but yields were low. It was an inefficient system and farmers were poor.
- The peasant farmers lived in hilltop villages and travelled daily to work on the latifundia.
- Only one-quarter of the farmers owned their own land.
- By 1950, 70 per cent of these farmers' land holdings were smaller than 3 hectares in size. To support their families they overworked the land, leading to overgrazing, overcultivation and eventually soil erosion.
- This system was called **minifundia**.

Land reform in the Mezzogiorno
- From 1950 onwards, most of the estates were bought by the state and the land was redistributed to the landless labourers.
- Holdings from 5 to 50 hectares were created.
- Farmers were trained to work their newly family-owned land efficiently, growing a mix of crops such as cereals, citrus fruit and traditional crops of olives and vines.

- Three related factors were put in place to support this new farming system:
  - an irrigation network to promote growth in summer
  - improved transport systems, such as autostrada to get high-value, perishable crops to market quickly
  - new villages and towns with all the important services such as schools, healthcare centres and leisure facilities were built.
- Today only one in ten of the region's workforce is involved in farming. The move to more intensive farming by the fewer farmers has also increased rural prosperity. The Mezzogiorno is now a leading supplier of citrus fruits, vegetables and olives to European markets.
- The most successful farming areas are on coastal lowlands and river valleys where irrigation water is available.
- The Metapontino is a coastal strip in the Gulf of Taranto. It was once a malarial swamp, but was drained as part of the land reform programme.
- Using the waters of the five rivers that cross the plain, irrigation produces cash crops such as citrus fruits, peaches, table grapes, strawberries, flowers and salad crops.

■ Secondary Activities

- By the 1950s only 17 per cent of Italy's workforce was located in the Mezzogiorno. Just as with agriculture, industrial development here has had its successes and failures.
- Government help was needed to encourage industrial development to take off. This help included:
  - generous grants and tax relief
  - major improvements in roads, autostrada, and modernisation of ports, such as Naples and Taranto
  - state-controlled companies having to make 80 per cent of new investment in the South
  - the creation of a number of key industrial development areas to act as a basis for regional growth.

Some results of the reforms

- Between 1960 and 2000 the region's industrial workforce almost tripled, to 1.4 million, and over 300,000 new jobs were created. This has reduced dependence on agriculture and increased the prosperity of the people.
- Almost 75 per cent of all new jobs have been in heavy industries such as steel, chemicals and engineering.
- Because the heavy industries are located on the coast, the inland rural areas have remained depressed.

- The most successful region is the **Bari–Brindisi–Taranto** triangle, where oil refining, chemicals and steel form the basis of this major industrial zone.
- The construction of a new deep-water port at Taranto was vital in the selection of this site for the country's largest iron and steel mill.
- **Latina–Fronsione** This is the fastest growing industrial area in the South. Over 250 new factories, including a car plant, employ over 16,000 workers.
- **Catania–Augusta–Siracusa** This region is one of the largest oil-refining, chemical and petrochemical complexes in Western Europe. Local deposits of potash, oil, natural gas and sulphur favour this type of industry.

■ Tertiary Activities

- Major investments were made in improving transport systems. The backbone of the system is the Autostrada del Sole, which runs from the Swiss border in the north of Italy to the 'toe' of Italy in Calabria. Another motorway runs along the east coast. These motorways create fast and efficient links between the north and south of the country.
- Port developments have improved access to the South.
- The long, hot, dry summers, dramatic coastal scenery, extensive beaches, and historic cities have much to offer the tourist. It also tends to be cheaper and less crowded than other Italian regions.
- Hotels have been improved to cater for the 12 million tourists who visit the south annually.
- More than 9 million of these tourists come from other parts of Italy.
- Sorrento, near Naples, is a major tourist centre.
- Located near Mount Vesuvius, Herculaneum and Pompeii, it is a busy coastal resort on cliffs that overlook the old town's fishing village and the Isle of Capri, another major tourist resort even in Roman times.

■ Human Processes

- Migration within Italy is a major factor, influencing population patterns inside the country. Between 1951 and 1971 over 4 million migrants left the South because of unemployment and poverty. Most of those who left were attracted to the cities of Milan, Turin and Genoa, or to the USA.
- Over 1 million people left the Mezzogiorno in the 1980s, and this out-migration trend continued into the 1990s.
- Since the 1990s increasing numbers of migrants from other countries, such as Albania and the former Yugoslavia, have been attracted to Italy. These migrants have not been welcomed by the Italian government.

> **STUDY NORRLAND: pages 96–101**
> *OR* **THE PARIS BASIN: pages 101–105**

NORRLAND:

A Cold Peripheral Region

The people of Norrland have interacted with the natural physical resources of the region to create jobs, income and a decent living standard for themselves and their families.

■ **Relief Drainage and Soils**

- Norrland lies on the eastern slopes of the Scandinavian Highlands that formed when the American plate and the Eurasian plate collided some 450 million years ago. It was then part of the Caledonian Fold Mountains. It is formed of some of the oldest and most weather-resistant igneous rocks, such as granite.
- These rocks formed when the ocean floor in between dipped under the Eurasian plate and melted to form masses of magma, which cooled deep within the folded mountains. Millions of years of weathering have now exposed many of these igneous rocks at the surface to form a plateau which slopes from west to east to the Gulf of Bothnia.
- Most of the rivers of this region flow in a south-easterly direction and cut steep-sided, parallel valleys into the plateau.
- There are many waterfalls where the rivers cross rocks of differing hardness, such as on the Angerman and Dal rivers. These locations are important for the production of hydroelectricity.
- The present surface was shaped when great masses of ice covered this region during past ice ages.
- Glacial erosion created many ribbon lakes, especially in the valleys of the largest rivers.
- As the ice sheets melted large amounts of glacial sands, gravels and boulder clays were deposited over the landscape, especially in the valleys. This material blocked drainage, creating numerous lakes that dot the region in the poorly drained areas.

■ **Climate**

The climate of Norrland is classed as Boreal. It has long, cold winters and short, warm summers. The climate is influenced by three factors:
- northerly latitude
- Continental influence
- rain shadow of the Scandinavian Highlands.

Northerly latitude

- Norrland stretches from 60° North to well inside the Arctic Circle. This creates large differences between winter and summer.
- In winter the movement of the sun to the southern hemisphere means long nights and very little daylight. Areas within the Arctic Circle have 24 hours of darkness and no daylight on 22 December. During winter, temperatures drop well below freezing point for up to 6 months. The average

▲ Norway and Sweden, showing ore fields and hydroelectric stations

Mark in the following towns where pulp mills and saw mills are located:

Inlea, Skellefta, Umea, Sundsvall, Gavle.

temperature for Gallivare, where iron ore is mined, in **January is –12°C.** Average temperatures are below freezing from October to March. The low angle of the sun's rays gives little heat at this time.

- In summer temperatures are much higher. The sun's rays then strike the ground at a greater angle and so create more heat. They also create more light, and areas within the Arctic Circle have 24 hours of daylight on 21 June. **July** temperatures are surprisingly high, with an average of **20°C.**

Continental influence

- Norrland is bitterly cold in winter because it is cut off from the warming influence of the North Atlantic by the Scandinavian Highlands. It is far from the sea.

- Regions that are far from the sea in winter get very cold. It is also attached to the large continents of Europe and Asia, which are bitterly cold in winter. This also helps to reduce temperatures at that time of year.
- Because it is so cold, the **Gulf of Bothnia freezes** over and only icebreakers can reach the northerly ports.
- In summer the cooling influence of the North Atlantic does not affect the region, as it does with Norway. Because temperatures in Continental Europe become very high, summer temperatures in Norrland are also surprisingly high for its latitude.

Rain shadow

- Norrland is to the east of the Scandinavian mountains.
- As the rain-bearing, south-westerly prevailing winds blow over these highlands they lose much of their moisture, 2,000 mm annually, on the Norwegian side.
- When they descend on Norrland's side they are drier, and annual rainfall totals only about 400 mm.
- So it is in the rain-shadow area and the region has a reasonably dry climate.
- **Most of its rainfall comes in summer.** Winter precipitation falls as snow.

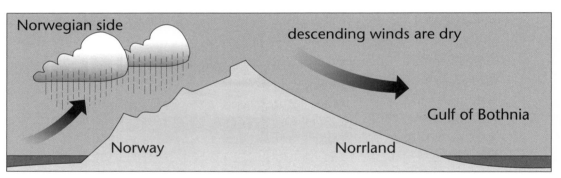

▲ Rain-shadow effect

■ Primary Activities

Norrland is less prosperous than most peripheral regions in Europe. This is because its people have used the local natural resources, such as forestry, water power for hydroelectricity and minerals to their advantage.

Farming

- The thin, infertile, acid soils, rugged landscape and short growing season make profitable farming almost impossible. Farming is confined to the narrow coastal plains and narrow river valleys such as the Angerman and Indals.

- Only 10 per cent of the land is given to crops and the region is not suited to cereals. Dairying is the most common type of farming, even though housing cattle throughout the long winters is expensive.
- Few people work in farming, but those who do give much of their land to forestry, or else they work part-time outside their farm to earn a living.

Forestry

- Norrland is part of the coniferous forest belt that runs across northern Europe, Asia and America. It is the chief timber reserve of the EU.
- Conifers such as pine and spruce are well adapted to withstand the long, bitter winters and the short growing season.
  - Conifer needles reduce moisture loss so the trees can survive through winter.
  - Conifers require few nutrients to grow and they grow quickly when compared to deciduous trees.
  - They are shallow rooted and grow well on thin, glacial soils.
- The forests are densest in the south, and the Angerman river is usually the northern limit for intensive commercial forestry.
- Forty per cent more trees are planted than are cut down to ensure the long-term future of the industry.

Mining

- Norrland has substantial quantities of varied mineral ores. There are large deposits of ores of copper, lead and zinc, which are mined at **Boliden** and moved by rail to smelters at Skelleftea. Local HEP is essential for the refining process.
- Further north at **Kiruna** and **Gallivare** there are vast reserves of iron: 3,000 million tonnes of high-grade magnetite ore.
- Most of the ore is exported through the ice-free port of **Narvik** in Norway, as ports are frozen during winter in the Gulf of Bothnia.

Hydroelectric power

- The heavy and constant rain in the Scandinavian mountains provides a plentiful supply of water for the many rivers that flow into the Gulf of Bothnia. A large number of hydroelectric power stations have been built along these rivers, which include the **Indals, Angerman and Lule rivers**.
- About half of Sweden's energy comes from water power. Over three-quarters of the supply comes from Norrland and huge electricity cables have been built to bring much of it south to the richer central lowlands.
- This large supply of hydropower has allowed:
  - a large population scattered over the region to live in extremely cold conditions;

- large-scale mining of mineral ores to take place;
- industries such as wood-pulp processing, steelworks and smelter plants to operate.

■ Secondary Activities

- Norrland has few advantages to attract industries other than those associated with forestry and mining.
- As industries modernised their operations, they reduced their workforce to remain competitive. Since this process also happened in forestry and mining, levels of unemployment have risen in the region.
- Wood processing is one of Norrland's major industries. Over 85 per cent of local wood supplies are carried by road, rather than being floated downstream by river.
- Many of the smaller, less efficient mills have closed and production is now concentrated on fewer, more efficient mills.
- Most of the higher-value paper mills are located in southern Sweden, owing to the advantages of road haulage and ice-free ports throughout the year.

■ Tertiary Activities

- Services are well developed because Sweden is committed to providing high levels of social welfare for its people. However, because of its considerable distance from the core region around Stockholm, Norrland's range of higher services is limited.

Transport

- Because of its isolated location and severe winters, long-distance road links are disrupted by snow and icy conditions.
- Its rail network is well developed, but its ferry services stop in winter because the ports are ice-bound.
- The railways provide an essential service of transporting mineral ores from both Lappland iron ore fields to Lulea and Norway's ice-free port of Narvik.

Tourism

- Tourism in Norrland is underdeveloped. The wilderness areas and forests are important nature reserves. Large areas are protected as national parks and they have great tourist potential.

■ Human Processes

- Norrland has a population of only 1.2 million people and a density of 2.1 people per sq. km. About 80 per cent of people live in the urban areas that have built up around the mines and sawmills.
- These towns are mainly along the coast and major river valleys.
- Most of the region is uninhabited and has extensive tracts of forest and wilderness.
- Although government incentives support development, the region's peripheral location, extreme climate and low population density make it unattractive to many people.
- Norrland has an **Objective 1 status**. So it receives structural funds to develop its infrastructure and protect its fragile environment.
- The majority of the people belong to the Lutheran Church.
- Some of the population belong to the distinctive cultural group called **Sami** who practise a more traditional religion. They speak a language called **Sapmi**, but Swedish is the dominant language of the region.

OR

THE PARIS BASIN:

A Core Region

■ Relief Drainage and Soil

- The Paris Basin occupies 25 per cent of France and so by itself is larger than Ireland.
- Structurally it is a broad, shallow **downfold** (depression) consisting of layers of sedimentary rock, one inside the other, and might be described as a series of stacked saucers.
- In the centre are sandstones and limestones, which are surrounded by belts of chalk and clay.
- In the east and south-east the edges of the chalk and limestone stand out prominently.
- Erosion has exposed the sedimentary rocks and has produced a landscape of alternate scarps (Côte de Meuse, Côte Moselle) and vales (Dry and Wet Champagne).
- The Paris Basin has a covering of **limon**, a wind-blown soil deposited here after the Ice Age.
- Limon soils are fine-grained, rich in minerals, and level.
- Champagne Pouilleuse (Dry Champagne) is named after the permeable nature of the chalk.

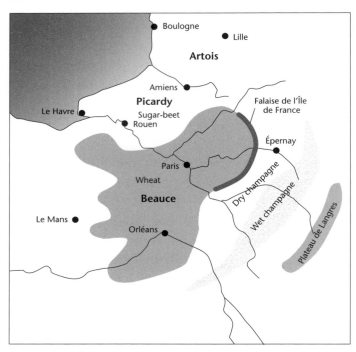

◀ The Paris Basin

- East of this again is an outcrop of clay, once an area of marsh, shallow lakes and damp soils, called Champagne Humide (Wet Champagne).
- The Paris Basin is drained by the Seine and its many tributaries (e.g. the Rivers Oise, Marne, Aube, and Yonne).

■ Climate

The Paris Basin includes two types of climate

- Along the English Channel (coastal strip) there is a maritime climate. South-west anti-trade winds blowing over the English Channel and the North Atlantic Ocean bring rain throughout the year.
- These warm, moist winds keep the climate mild in winter and warm in summer.
- Because the Paris Basin forms part of Continental Europe, its summer temperatures are higher than average for its 49° latitude.
- Inland, towards Paris and the scarplands, a transitional type of climate is experienced.
- This means that it forms a middle zone between the maritime climate on the coast and the continental type in central Europe.
- So Paris has average temperatures of 2.5°C in January and 18.6°C in July. Its annual rainfall is 570 mm.
- So it has a cool, dry winter and hot summers, with a rainfall maximum in spring and summer.

■ **Primary Activities**

Agriculture

- The rolling expanses of loamy soils grow wheat and sugar-beet on the Île de France. In Beauce, large farms and level, open landscape allow intensive mechanised cultivation of cereals: wheat, barley, and sugar-beet.
- Many farms exceed 80 hectares (200 acres) and have large fields where a high degree of mechanisation is achieved. The area is called the granary of France.
- Falaise de l'Île de France, a limestone scarp, marks the northern limit of vines in France.
- The scarp slopes are sheltered and sunny, so this is a region of vineyards famous for the sparkling white wine known as champagne.
- Reims and Epernay are market centres for the wine industry.
- Mixed farming is carried on in the Dry Champagne, with animal farming and cereals such as wheat. In Wet Champagne dairy farming is practised on the valley floor.
- Surrounding Paris, market gardening is practised on an intensive scale to supply the conurbation with a large variety of vegetables.
- The population of Paris is approximately 10 million.

■ **Secondary Activities**

Industry

- Paris is a focus of routes – air, water, road and rail – on a bridging point on the River Seine.
- It is the centre of an important inland waterway system, and is connected by canal and canalised rivers to the Plain of Alsace, the Rhine, the Loire and the Saône, and to the sea at Le Havre.
- Although 160 km from the sea, Paris is an extremely busy inland port.
- Products from the Nord and the Paris Basin, as well as imports, are stored on the quays upriver from Le Havre and Rouen.
- Paris has 20 per cent of the national workforce and a highly diversified economy: craft industries, such as perfumes and fashion clothing, vehicle assembly, oil refining, and chemicals.
- Industry in Paris has been successful because:
  - (a) it is in the centre of a rich agricultural hinterland, with processing such as milling and canning;
  - (b) it is the centre of French rail and road networks. So it is the ideal location for assembly-type manufacturing as components can be brought from all parts of France; and
  - (c) it is built on a wide and deep river, the Seine, and has excellent dock facilities for the export of products and the import of raw materials.

- Le Havre is an important port at the estuary of the Seine. It has many industries, such as oil refining and chemicals, which occupy a large area on the northern bank of the river.
- Ship repairs and maintenance are also important in Le Havre.
- Rouen on the River Seine is an inland port with large oil-refining and chemical industries.

■ Tertiary Activities

Transport

- Paris is the centre of the French transport system.
- All routes meet in Paris and the Autoroutes and rail networks connect Paris to other regions. Because the landscape of France is low-lying and undulating it was easy to develop this network.
- Paris is the hub of the TGV rail system. The TGV is a high-speed train network on which trains can reach speeds of 300 km per hour.
- Paris is also connected by the TGV system to London, via the Channel Tunnel, and to Brussels.
- There are about 350 TGV trains in operation, making it a very efficient system.
- Transport systems within Paris create thousands of jobs. Part of this service is the Métro, which is the underground rail system in the city.
- The Métro is connected to the national network, the SNCF, and the Reseau Express Régional, an underground rail system that extends into the suburbs.

Tourism

- Millions of tourists visit the city annually.
- Its wide boulevards that meet at the Arc de Triomphe, ornate buildings, art galleries such as the Louvre, monuments such as the Eiffel Tower and street attractions such as Montmartre on Sundays all add character to the city. Euro-Disneyland is located 30 km to the east of the city.
- To the west of the city is the Palace of Versailles, the former royal palace of Queen Marie Antoinette and King Louis XVI.
- The Paris Basin has many other attractions, such as the cathedrals of Notre Dame, Chartres and Reims. Paris has been described as the most beautiful city in the world.
- One of the most successful developments in Paris has been at La Défense, to the west of the city centre.
- Large office complexes have been built creating thousands of office jobs in the tertiary sector.

■ **Human Processes**

- The total population of the Paris Basin is 21 million people. This is more than one-third of the population of France.
- The Île de France, the centre of the Paris Basin, has 11 million people. This creates high densities in the centre, in stark contrast to the edges of the basin where densities are much lower.
- The birth rate for the Île is 15 per 1,000, because a large proportion of the population is young. The death rate here is only 7 per 1,000, so natural increase is 0.8 per cent.
- The edges of the Paris Basin are suffering from out-migration and these migrants are moving towards Paris because of its attracting forces.

Racial composition

- A shortage of labour in cities along the River Seine after World War II attracted migrant workers from Spain and Italy. As the Spanish and Italian economies picked up and the flow of workers from these countries slowed, Portuguese and Algerian migrants came to France in large numbers, along with migrants from other North African countries.
- About 1.2 million foreign workers live in the Paris area.
- Some North African countries were colonies of France and so their people were entitled to enter France.
- Large numbers of migrants live in inner-city ghettos or poor suburban neighbourhoods.
- Unemployment in France has led to racial tensions in the past.
- The rise of Islamic fundamentalism has led some Islamic women to adopt more traditional codes of dress, and this has resulted in some tension within schools and with school authorities.

CORE TOPIC 15
A SUBCONTINENTAL REGION

STUDY **ONE** SUB CONTINENTAL REGION:
INDIA: pages 106–114
OR
THE AMERICAN SOUTH-WEST: pages 114–119

INDIA

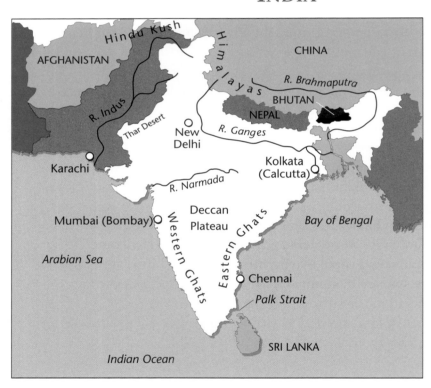

◀ The Indian subcontinent

- India is a subcontinental region. It is one of the most heavily populated and also one of the poorest regions in the world.
- The physical environment and human activities have a profound effect on each other.

■ **Relief, Drainage and Soils**

There are three main physical regions in India:

- Northern Mountains
- Indus–Ganges Plain
- Southern Plateaux.

Northern mountains

- Extremely high mountains in the north of the country separate India from its neighbours.
- They extend from the Hindu Kush in the north-west, through the Himalayan range to the extreme north-east of the country.
- Mount Everest, the highest mountain on Earth, and the next 23 highest peaks are all in this range.
- They were formed by the collision of two of the Earth's crustal plates, the Eurasian plate and the Indian plate.
- This collision compressed the Earth's crust and buckled it upwards to form these fold mountains.

Indus–Ganges plain

- The Indus–Ganges Plain is a huge depression or syncline that formed south of the mountain chain. It follows the Indus river valley from Pakistan through the Ganges valley, and ends with the double delta of the Ganges and Brahmaputra rivers in Bangladesh.
- It is covered with thousands of metres of alluvial (river) soils that have been washed into the depression by India's three most important rivers – the Indus, the Ganges, the Brahmaputra – and their tributaries.
- About half of India's people live in this region.
- The rivers are swollen in summer by meltwaters from glaciers and monsoon rains from the surrounding mountains.
- Extensive areas of lowland are flooded by these waters, which deposit highly fertile soils on their flood plains.

Southern plateaux

- Southern India is made up of a number of plateaus. The Deccan plateau, the largest, is tilted from west to east.
- Two mountain ranges, the Western Ghats and Eastern Ghats, border narrow coastal lowlands.
- Both of these have an effect on onshore winds and rainfall amounts for peninsular India.

■ **Climate**

- The climate of India is Tropical Continental Monsoon. Most of India is in the tropics.
- Only the mountains of the north and north-west have frost. Temperatures year-round are high.
- India's climate can be divided into two main seasons:
    - the dry monsoon
    - the wet monsoon.

The Dry Monsoon

- This occurs from October to February, when cold winds blow outwards from a high-pressure area in the centre of Asia. They are land winds, so they are dry. They bring freezing temperatures and snow to the mountains in the north.
- From March to June these winds become warmer, and by June temperatures can be as high as 49° C in the Ganges valley.

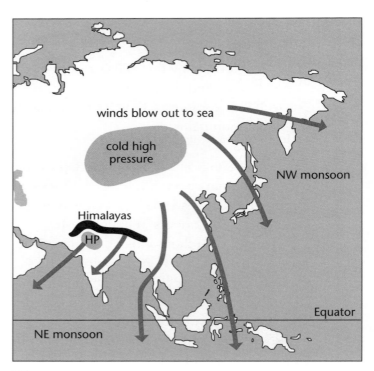

winds blow out to sea

cold high pressure

NW monsoon

Himalayas

HP

Equator

NE monsoon

▲ October to June monsoon season

The Wet Monsoon

- From mid-June to September, warm ocean winds are sucked into a low-pressure area in the continent. One wind blows as a south-west monsoon from the Arabian Sea. Some of this air is forced to rise over the Western Ghats and intense rain falls. The second wind blows from the Bay of Bengal and veers north to blow along the Ganges and Brahmaputra valleys. Up to 10,000 mm can fall in some areas in a six-week period.

- As the winds move west along the Ganges valley, rainfall reduces. When they reach the extreme north-west the winds have become dry, leading to desert conditions.
- The monsoons bring essential water supplies to India's population. If they don't arrive it can mean widespread famine in the country.

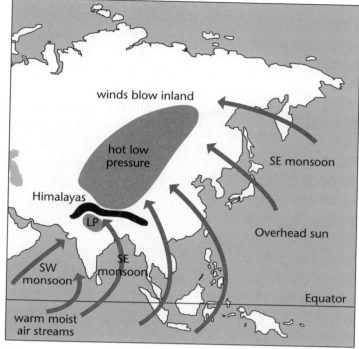

▲ June to September monsoon season

■ Primary Activities

Farming

- India's cultivated land is equal in area to the total cultivated land of the EU countries. Arable farming, especially cereals, is the main type of farming.
- Two-thirds of India's one billion people depend directly on the land for their living.
- Almost half of rural families have farms of less than 0.5 hectares, or no land at all.
- A quarter of India's agricultural land is owned by less than 5 per cent of farm families. Most family farms are broken up into tiny, scattered plots.
- Farming is mainly intensive subsistence. This means people depend on their own food supplies to feed themselves.
- Rice is the chief crop. Cereals such as wheat and millet are grown in drier areas.
- Almost all planting, weeding and harvesting is done by hand.
- Double-cropping is practised. While rice is grown in the wet season, other crops such as cereals are grown in the drier season.
- The country's rapidly growing population places a huge demand on annual output.

- Genetically modified, high-yield varieties of rice and wheat have been introduced and are now grown. These varieties are resistant to many diseases and pests.
- This practice is called the Green Revolution and has led to India being a net food exporter.
- India has the largest livestock population in the world. Many are in poor physical condition. The slaughter of cattle is illegal in many states because of a Hindu religious belief that the cow is a sacred animal.
- Most of the beef that is eaten comes from cattle that have died of old age. Many cattle are malnourished and old livestock are allowed to roam as strays or may be sent to special compounds until they die.

Mining

- India has large reserves of iron ore and copper. Other mineral ores are bauxite, from which aluminium is made, and zinc, gold and silver.
- Most of India's oil comes from Mumbai (Bombay) High Field in the Arabian Sea.
- The most important coal-producing regions are Bihar and West Bengal.

■ Secondary Activities

- After independence in 1947, only 2 per cent of the labour force was employed in industry.
- Industry was concentrated in the major cities of Mumbai (Bombay), Kolkata (Calcutta) and Chennai (Madras).
- Indian industry has three factors in its favour:
  - a large home market
  - a wide range of natural resources, such as coal and iron ore and
  - a cheap labour force.

The government has focused on new industries such as:

1. Agri-industries: the manufacture of fertilisers, machinery and food processing to benefit rural communities.
2. Consumer goods industries and small-scale, labour-intensive craft industries. These employ a lot of workers and together with traditional skills they could be competitive on export markets.
3. Community-based developments and self-help schemes in rural regions: this was to create jobs in rural regions where over 70 per cent of the people lived. This would prevent, or reduce, rural–urban migration.
4. High-tech industries: the growing educated workforce is attracting computer software companies to India. India produces more university graduates than the

USA and Canada combined, and 40 per cent of them are in science and engineering. Most of these new industries are located in urban regions such as Kolkata, Mumbai and Chennai and their hinterlands.

- Following independence, national planning recognised the importance of urban centres for economic development.
- A new capital, New Delhi, was set up.
- Major urban growth centres were also established, based on existing cities.

  - **Mumbai (Bombay)**

  Mumbai has attracted growth industries such as electronics and pharmaceuticals. It also has traditional industries like food processing and textiles.

  - **Chennai (Madras)**

  Chennai forms the core of this southern industrial zone. Textiles and light engineering are important industries. Many multinational computer software companies have set up here. The region is called India's 'Silicon Valley'.

  - **Kolkata (Calcutta)**

  Heavy industries such as iron and steel are long established, owing to local sources of coal and iron ore. The Indian-owned Tata Iron and Steel Company is one of the largest in the world, with branch plants in South East Asia, the Middle East and South Africa. Traditional industries such as cotton, clothing and jute manufacturing are here also.

■ Tertiary Activities

Services

- India's service sector is underdeveloped. So many of India's population are poor and do not have the money for education, healthcare, or much else, even if these services were available. As with similar economies, there are two levels of services.
- One type caters for what are regarded as the rich; for them there is the full range of services.
- The other type caters for the poor, or underclass. It is similar to what you would find in any large city, and is the informal sector. There are the unlicensed street sellers such as shoeshine people and street vendors, alongside illegal activities such as prostitution and drug dealing.

Transport

In 2000, half of India's villages did not have access to tarred roads suitable for vehicles. These communities use dirt-track roads and carts drawn by cattle.

Tourism
- India's varied landscape, its history and its natural wonders offer vast potential for the tourist industry. Attractions include:
  - the Himalayan mountains;
  - the many palaces and religious temples of the Hindu, Buddhist, Sikh, Jain and Muslim religions;
  - the physical landforms of its major rivers
  - a wide variety of wildlife.
- Tourism is on the increase. However, much remains to be done, especially when there is such pressure of population numbers. The obvious poverty of many of the people can be upsetting for the visitor.

■ Human Processes

Population
- India's population is greater than one billion people. With a natural increase of 1.6 per cent annually, its population increases by 16 million people every year. Its population could reach 2 billion by 2040. This creates difficulties for the Indian government, such as ensuring a sufficient food supply, sufficient jobs and controlling rural–urban migration.
- The country has only recently entered the third stage of the population cycle.
- Even though healthcare has improved and the death rate has reduced, death rates are still high.
- Because rural families are large, it is difficult to control population.
- Large families are seen as a positive aspect of life, rather than as a burden to feed and clothe.
- Because India has a very young population, it will continue to have a large natural increase for the near future.
- India's population is very unevenly distributed. High population densities are located in the Ganges valley, along coastal lowlands and in cities and their hinterlands. The interior regions have low population densities.

■ Culture

India has many different culture groups. A number of outside factors have complicated matters. These include:
- the migration of Indo-Europeans
- the spread of Islam
- British occupation.

- The people of India speak many different languages: there are over 1,600 different languages and dialects. Schools teach in 58 different languages.
- National newspapers are published in 87 languages and radio programmes are broadcast in 71 different languages. This creates difficulties and disunity between culture groups.
- The two main language families in India are Indo-European and Dravidian.
- Hindi is the official state language, but its position of importance is resented by other language groups.

Religion

Hinduism

- Hinduism is the dominant religion in India. It creates a multi-layered society in which people are divided according to class or **caste**. At the top of society are the priests or Brahmins, and other high-ranking people such as officials or professionals. At the bottom are the lowest castes who do the menial or dirty work. They are believed to be unclean or 'untouchables'.
- Belonging to a caste is decided by birth and one cannot move up the system in a single lifetime. Caste members may only socialise or marry within their own group. However, some movement occurs within castes in urban areas, as traditional values are not so strong there. But still many untouchables are restricted to their own caste and have little chance of improving their lot.
- The Hindus regard the cow as sacred, so it cannot be killed. Consequently millions of undernourished cattle roam the streets and countryside and are a drain on resources. Cattle provide milk, pull carts and their dried dung is used as fuel.
- While Hinduism is the dominant religion, other religions are important. These include:

Islam

- There are about 200 million Muslims in India. This religion was introduced through trade and it is most common in the Indus and Ganges basin.
- It is not common in peninsular India. Islam accepts all converts as equal and rejects the caste system, so it became attractive to many Indians.

Sikhism

- This religion was founded in the 15th century. It does not have a caste system either. The Sikhs are a powerful cultural group and are centred in the Punjab, an important farming region.

Buddhism and Christianity

- Buddhism and Christianity are two other minority religions in India.

■ **Human Processes**

The political–religious divide in India

- India was a colony of Britain. After independence in 1947 it was divided into two states: India, a Hindu state, and Pakistan, a Muslim or Islamic state. This division was based on religious grounds and caused many minority religious groups to remain within the Indian state.
- Large-scale migration resulted, due to fears of persecution. Over 15 million moved home. Many Muslims left India for Pakistan, and many Hindus left Pakistan for India.
- Pakistan was divided into two parts in the beginning: one to the west, in the Indus valley, and called West Pakistan; and the other, East Pakistan, in the Ganges valley. They were separated by a long distance, with northern India in between.
- This did not work out. So East Pakistan broke away from West Pakistan and became **Bangladesh**. Some territory is still in dispute between Pakistan and India; this disputed region is called **Kashmir**.

OR

THE AMERICAN SOUTH WEST

The American South West is a region that borders Mexico, stretching from California in the west to Texas in the south. Its states include **California, Nevada, Arizona, New Mexico** and **Texas**.

■ **Relief, Drainage and Soils**

The physical landscape of the American South West varies enormously.

- To the west in California is the Central Valley that separates the **Coastal Range** from the **Sierra Nevada Mountains.**
- Further east are the highlands and **inter-montane plateaus and basins** of the mountain states of Nevada, Utah and Arizona, the largest of which is the Great Basin of Nevada.
- Further east still are the **Rocky Mountains**, which fall onto the High Plains of Texas.
- The Great Basin is the largest basin of internal drainage in the USA.
- The Great Salt Lake, near Salt Lake City, was formed from salts that resulted from the evaporation of river waters flowing towards this depression.
- The Colorado and the Rio Grande are the two largest rivers in the region.

▶ Western America and Mexico

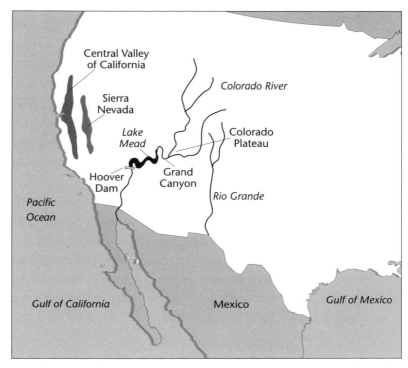

- The Colorado flows through the Grand Canyon gorge, which formed as the river eroded vertically through this barren landscape in northern Arizona.
- The Rio Grande separates Texas from Mexico.
- The soils in the lower stage of the Rio Grande are rich in alluvial deposits. The coastal stretch of Texas also has rich alluvial soils.

■ Climate

- Much of the South West is **arid or semi-arid**, with average annual precipitation below 250 mm. Southern California, Arizona and New Mexico are either desert or semi-desert.
- **Death Valley** in southern California, which is below sea level, has recorded the hottest temperatures in the western hemisphere.
- Central Valley in California has a Mediterranean climate of hot summers and warm winters with temperatures somewhat like our summer days.
- The factors that affect climate, such as distance from the sea and altitude, combine to create distinct climates in some areas.
- They vary from sub-tropical along the Gulf Coast to Mountain climate in the Sierra Nevada and Rockies, and cover a vast region that stretches 3,000 km from coastal Texas to California.
- The region forms part of the Sunbelt of the USA. Many regions that border Mexico have over 2,500 hours of sunshine annually.

■ **Primary Activities**

Farming

- The availability of water supplies and altitude influence agriculture in the region. Many family farms exist here.
- The industry is dominated by large farm units, in many cases owned by **agri-companies** which include major food companies involved in the production, processing and distribution of food products.
- Agriculture is very market-focused and highly industrialised, with heavy investment in machinery.
- The greater proportion of the region is dominated by cattle ranching in the dry lands of Nevada, Utah, New Mexico and Arizona. Many farm workers use cowboy skills on these enormous ranches.
- Rainfall is uncertain and the low stocking rates means that cattle numbers per hectare are few.
- This is an extensive form of farming and is very different from the intensive form of agriculture practised in the Central Valley in California.
- Irrigation is widely used for intensive farming in response to the rapid rise in population of this region over the past 40 years.
- About one-third of the fruit crops and one-third of the truck crops of the entire United States are produced in California.
- One of the largest networks of irrigation systems in the world supplies water to the farms in California.
- The farms specialise in **fruit and vegetables**.
- Citrus fruits are widely grown and California is the second largest producer in the United States, after Florida. The fruits that are grown include dates, peaches, grapes, cherries, tomatoes and plums.
- In California there are many regions that specialise in a single crop, such as the Napa and Sonoma valleys which specialise in vines and the wine industry.
- Much of the agricultural labour is carried out by migrant Mexican labourers.

Mining

- In the past many people were attracted to the South West because of its minerals, such as gold and silver.
- The discovery of gold in 1848 led to the **Californian Gold Rush** and in 1859 the Comstock Lode, a silver deposit, led to a large number of settlers to the Great Basin of Nevada.
- Today, **oil and natural gas deposits** are the main employment providers. Texas, in particular, has enormous deposits.

- The wealth of the region fluctuates and depends on the value of oil. During periods of high oil prices profits soar and employment improves, while when oil prices are low the level of employment falls.
- Mining is still important. Uranium, sulphur, gold and copper, lead and zinc are all mined.
- The mineral deposits owe much of their origin to igneous activity associated with subduction of the Pacific plate under the North American plate.
- **This region is part of the leading edge of the American Plate.**

■ **Secondary Activities**

- In recent years manufacturing has moved away from the north-east USA to southern California and Texas, where unskilled labour is cheaper. This is because of the large numbers of Mexican migrants who have settled in towns along the Mexican–USA border region.
- The advantages of this region are as follows.
    - There is plenty of cheap land for factories.
    - On the Pacific Rim it is close to Asian markets, and close to the Atlantic Ocean for access to European markets.
    - there are many industrial raw materials in the area, such as oil and gas for the petrochemical industry.
    - Cities do not suffer from urban decay and the quality of life is good.
    - The US Air Force bases in the region have given rise to many factories making military equipment.
    - it is well connected by road to markets in the USA.

Manufacturing in Texas

- Texas is eight times the size of Ireland. It is the most important state in the USA for the production of the oil and gas that form the basis of many industries.
- The coastal cities on the Gulf are known as 'the chemical crescent'.
- There are 30 oil refineries on the coast, as well as numerous petrochemical works and fertiliser plants. Offshore oil and gas platforms are located off the coast.
- The aerospace industry is centred on Houston – so aircraft, space-related technology and satellite equipment are manufactured in the region.
- Much industry is located in the industrial triangle of Houston, Dallas–Fort Worth and Austin–San Antonio.
- Austin, the home of Michael Dell, is the corporate headquarters of the Dell Corporation which manufactures PCs.

Silicon Valley in California

- Silicon Valley is the best-known centre for information technology in the Western world. Famous companies, including Hewlett-Packard, Intel, Sun Microsystems, Apple, and IBM, are all located here. Silicon Valley is located south of San Francisco.
- Stanford University and the University of California are central to the growth of Silicon Valley.
- Approximately 50 per cent of all high-tech equipment in the USA is manufactured here.

Manufacturing on the Mexican border

- Multinational companies locate twin manufacturing plants on each side of the Mexican–American border.
- The Mexican companies supply products that are made with low labour costs to their American companies, which then export the finished product.
- These manufacturing centres on the Mexican side are called **maquiladoras**, and many of them have experienced rapid growth in recent years.
- In 2000, 23 per cent of all Mexican manufacturing production came from maquiladoras.

■ Tertiary Activities

- Tourism is a major industry in the South West. The Grand Canyon, Zion National Park, Yosemite, and Carlsbad Caverns, one of the largest limestone caverns in the world, are all located here.
- Many cities, such as San Francisco, Las Vegas, Hollywood in Los Angeles, are also major tourist attractions.
- Las Vegas is the gambling capital of the world.
- It is a city built especially for entertainment, in a desert landscape.
- San Francisco is a major tourist town. Fisherman's Wharf, the prison of Alcatraz on an island in San Francisco Harbour, the trams on Powell Street and its Chinatown attract many visitors.
- Public transport is a problem for Los Angeles.
- The lack of an underground system, as well as few bus routes, have led to an extreme form of car culture where everybody has a car – and needs one if they must travel.
- With over 8 million cars, Los Angeles has the most polluted air of any city in the USA. Recent steps have been taken to reduce pollution.

■ **Human Processes**

- Total population for the region is about 60 million people. Because many cities are affected by urban sprawl, large numbers of people have difficulty travelling to work owing to traffic problems.
- The South West is a multi-cultural society. Only a small population of Native American Indians live on reservations. Some have become farmers, stockmen and truckers.
- There is a large Hispanic community in the South West. The term Hispanic refers to people of the Spanish-speaking world who come from Mexico and other countries to the south, in Central and South America.
- Even though there are Hispanics who have become wealthy in the USA, the majority stay in poverty for many years after arriving. Numerous television channels are broadcast in Spanish.
- Many Hispanics live illegally in the USA.
- A large number of Asian ethnic groups live in the South West. They include Chinese, Japanese and Vietnamese.
- The majority of these people live in California, the American region nearest to where they came from originally.

CORE TOPIC 16
ECONOMIC, CULTURAL AND POLITICAL PROCESSES INTERACT WITHIN REGIONS

CASE STUDY 1: NORTHERN IRELAND

■ **Historical Background**

- Scottish settlers were brought to the region during the Ulster Plantation. These settlers had a different culture, religion and traditions from the local Catholic population.
- In the nineteenth century, Ulster experienced the development of heavy industry based on coal supplies from Scotland.
- So when partition occurred in 1921 the Republic had an underdeveloped economy, while the north-west was a thriving industrial society.
- The majority of Irish people saw that political independence was essential for the economic and cultural development of the other three provinces, Connaught, Leinster and Munster.

■ **Changing Relationships on the Island of Ireland**

Economic trends

- Migration was dominant throughout the decades up to 1960.
- Since then the Republic has attracted foreign multinationals to Ireland.
- Ireland now has a developed, modern, industrial economy with high-tech industries that have created 'the Celtic Tiger'.
- Northern Ireland, on the other hand, has a depressed economy. Its old textile and shipbuilding industries have gone.
- Civil unrest and religious bitterness have discouraged foreign investors.

Political interaction

- Tensions between the Republic of Ireland and Northern Ireland faded somewhat between 1921 and the 1960s.
- The refusal of the British government to grant full civil rights to Catholics in the North led almost to a state of civil war that continued virtually continuously until the 1990s.
- The signing of the Belfast Agreement led to new political interactions based on the 'Strands'.
- These new political bodies are a new Northern Ireland Assembly, a North–South Ministerial Council and a British–Irish Inter-Government Conference.
- These bodies are designed to create inter-relationships and cooperation rather than division.

CASE STUDY 2: THE SAPMI REGION

- The Sami or Lapps are a cultural group who live in the Sapmi region, which includes the northern parts of Norway, Sweden and Finland.
- A total of about 60,000 Sami live in this region. They are culturally distinctive.
- They have their own language, music and crafts.
- The Sami are a minority within their own region because people from other parts of these countries have migrated here to mine iron ore and other minerals, as well as to work in forestry.
- Because of their Arctic environment, the Sami hunt, fish and herd reindeer. These are their traditional jobs. It is difficult for young Sami people to make a sufficient living from these occupations.
- The young are not attracted to these jobs. They prefer to work in mining and forestry, where they can make a decent living. Many others migrate to cities in the south of their countries.
- The Sami way of life depends on the nomadic use of large regions.

- Their activities conflict with mining or forestry groups.
- Even though Sami rights to traditional lands have been guaranteed, they have lost much land to forestry.
- Little support for the Sami culture comes from government agencies. However, the European Bureau of Lesser-used Languages (EBLUL), an organisation at the European Headquarters in Brussels, works on behalf of those who speak minority languages.
- This group creates policies to protect these languages from extinction. In this way the EU helps to protect the Sami people.

CORE TOPIC 17
THE INTERACTION OF CULTURAL GROUPS WITHIN POLITICAL REGIONS (COUNTRIES)

- Some minority culture groups with a strong self-identity feel their interests are not represented by the larger host country in which they live; so the links that tie these regions together become weakened.
- What emerges are nationalist groups that look for more powers of self-government. This is called **autonomy** or **devolution**.
- A more extreme agenda would involve a new and separate state from the majority population.

CASE STUDY: THE BASQUES

- The Basques are a cultural group who occupy an area on each side of the Pyrenees in Spain and France.
- The core region of the Basques is in northern Spain and has a population of 2 million people. **Bilbao** is its largest city.
- Of the four Basque districts in Spain, three of them form a political unit called **Euscadi**. It has its own president and parliament but is represented internationally by Spain.
- The Basques have lived in this region for over 4,000 years, so they have developed their own distinct cultural identity.
- Their language is called **'Euskera'** and it is unrelated to any other language.
- The Basques in northern Spain and south-west France have been linked to a militant group called ETA that seeks independence.
- ETA is a separatist terrorist group that has been responsible for many terrorist bombings and assassinations of political opponents in the Spanish government.

- They believe that complete independence from Spain and France can be achieved only by military means, similarly to what the IRA believed about the reunification of Ireland.
- Initially ETA was founded in 1958 because the Basque people were oppressed during the reign of the fascist dictator Franco in Spain.
- Because they were different he saw them as a threat, and he tried to eliminate all opposition to his rule.
- In the past the Euskera language was illegal and large numbers of Basque nationalists were put in prison for calling for independence and for their culture to be recognised.

▲ Spain showing the location of the Basque country

The 'zero tolerance' of terrorist organisations since 9/11 has hardened attitudes against terrorist groups such as ETA and the IRA.

- The influence of ETA on Basque society is not measured only by the impact of its armed struggle. A key aspiration of ETA is to restore the Basque region to its full cultural personality.
- Traditionally the Basques were herders. Today's Basque region has numerous businesses.
- Bilbao has many steel factories and shipbuilding yards, based on its local coal and iron-ore supplies.
- The region has a positive Science and Technology Plan to strengthen this type of business and industry.
- Basque cuisine is based on seafood, especially cod and hake.
- The running of the bulls in Pamplona is an annual event, when six bulls are allowed to run freely through the city's streets before being killed later that day by matadors in a bullfight.
- Catalonia is another Spanish region that has a distinct cultural identity.

CORE TOPIC 18
THE FUTURE OF THE EUROPEAN UNION

New developments within the EU will influence trade, politics and sovereignty issues.

NEW TREATIES

■ **The Maastricht Treaty, 1992**

This established the three 'pillars' of the EU:

– the economic pillar, which promoted economic development and so introduced the Single European Market (SEM). In 1993 the Single European Market allowed for the free movement of goods, services and people within the EU. The Economic and Monetary Union (EMU) and the new currency, the euro, were introduced in 2002.

– the political pillar, which recognised the need for the EU to strengthen its role in foreign affairs and defence. So the Common Foreign and Security Policy was introduced. In 2004, a new Constitution for the EU was agreed and the EU appointed a new ambassador, John Bruton, to the USA to improve relations in the aftermath of the war in Iraq.

– the social pillar, to give more social benefits to EU citizens and to deal with crime and immigration problems.

■ **The Amsterdam Treaty, 1997**

This placed employment and citizen rights at the 'heart' of the EU.

■ **The Nice Treaty, 2002**

This allowed for change in the institutions and voting systems so that the EU could be enlarged further.

SOVEREIGNTY

Each member had to give up some degree of independence or sovereignty, so four main political institutions were created to regulate the Union:

– The European Commission in Brussels
– The European Parliament in Strasbourg, Brussels and Luxembourg
– The Council of the European Union (Council of Ministers)
– The European Council.

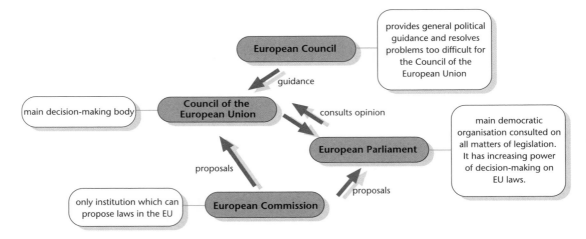

▲ How decisions are made in the EU

DEVELOPMENT AND EXPANSION OF THE EU

- In 1957 the Treaty of Rome created the European Economic Community (EEC). Its purpose was to increase trade between six countries in the core of Europe. Since then there have been five enlargements.
  - **1973: Denmark, Britain and Republic of Ireland were added**
  - **1986: Spain and Portugal added**
  - **1990: East Germany becomes part of Germany**
  - **1995: Austria, Sweden and Finland added**
  - **2004: the Eastern Bloc countries were added.**

CORE TOPIC 19

CHANGING BOUNDARIES IN LANGUAGE REGIONS

The size and shape of language regions can change over time.

THE IRISH LANGUAGE REGIONS

- There were 1.5 million Irish speakers in 1851. Most of those speakers lived in the western half of Ireland at that time.
- From 1861 the number of Irish speakers declined to 544,000. This was due to:
  - large-scale emigration
  - the growing popularity of English.

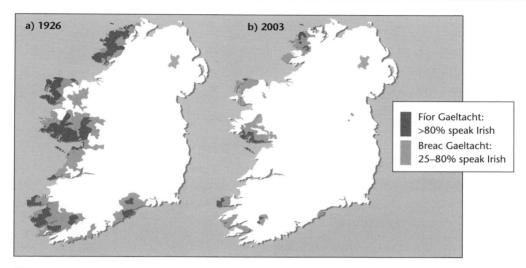

a) 1926   b) 2003

Fíor Gaeltacht:
>80% speak Irish
Breac Gaeltacht:
25–80% speak Irish

▲ Ireland's Irish-speaking regions, 1926 and 2003

■ **From 1926 to the Present**

- After independence the Irish government was committed to supporting Irish.
  – Irish became the official language of the state.
  – Irish was compulsory at school.
  – Official Gaeltacht regions were identified.

■ **At Present**

- About 1.5 million people can speak some Irish.
- The largest numbers are in Leinster. However, few are fluent.
- The Gaeltacht boundaries have diminished to tiny pockets that are located on peninsulas in the West of Ireland.

CORE TOPIC 20
Urban Growth and City Regions

■ **World Urban Growth**

- In 2004, half of the world's population of 6 billion people lived in cities.
- By 2025 about 80 per cent of people will live in cities.
- The growth rate of this trend is fastest in the developing world.

■ **In the European Union**

- About 80 per cent of the population lives in cities.
- Urban sprawl is a problem in every EU country.
- Rush hours create traffic jams in most city regions.

CASE STUDY: THE RANDSTAD, HOLLAND

- It is one of the most urbanised regions of the EU.
- The Randstad is shaped like a horseshoe.
- It is a polycentric city, i.e. many cities have expanded to create this urban area.
- There is a 'green heart' at its centre, with small villages and towns scattered within a rural landscape.

■ **Planning Solutions for the Randstad**

- Five regional centres have been chosen for new urban development to reduce pressure on the green heart.
- High-density developments will be included.
- Buffer zones will be used to control urban sprawl.
- Strict controls will be used to prevent cities joining up to create a megalopolis.

CASE STUDY: DUBLIN

It is Ireland's capital and it is a primate city because:
- It is the focus of the country's transport network.
- it is the biggest employment centre.
- There is a huge urban sprawl problem owing to the cost of land in the city.
- It relocated large numbers of inner-city residents into suburban housing estates.
- Three new towns were created around Dublin: Blanchardstown, Clondalkin and Tallaght.
- Many villages have increased in size to become small towns.
- Long journeys are made by commuters to Dublin, effectively extending Dublin's urban boundaries.

■ **Planning Solutions for Dublin**

Ireland has published its national strategy. This includes:
- large urban centres called **gateways** to reduce the dominance of Dublin
- smaller centres called **hubs** to draw some development away from gateways
- new routes called **strategic road corridors** to create links between gateways, hubs and Dublin.
- redrawing of local authority boundaries. Four new councils were created: Fingal, South Dublin, Dun Laoghaire–Rathdown and Dublin Corporation.

CORE TOPIC 21

CHANGING POLITICAL BOUNDARIES AND CULTURAL GROUPS

Changes in political boundaries can have an important affect on cultural groups.

Some people, as a result of changes in a country's boundaries, find themselves living under a different government or political system.

CASE STUDY: THE PROBLEM OF KASHMIR – A RELIGIOUS CONFLICT

- Remember your study of how India, Pakistan and Bangladesh became independent countries as a consequence of partition in India after independence. Different religious beliefs was the main reason for this partition.
- Once independence was given, the ruler of each state had to decide whether to join India if there was a Hindu majority, or Pakistan if there was a Muslim majority.
- Violence broke out in the Kashmir valley between the minority Hindu population, who looked to India for support, and the majority Muslim population, who looked to Pakistan for support.

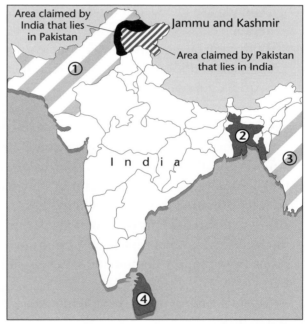

▲ India and surrounding countries. Name the countries 1–4.

- War broke out when the ruler opted to join India.
- Pakistan claimed Kashmir because there was a majority Muslim population. India claimed it because the ruler had decided to join India. Today the region is divided into two parts.
- The part nearest India is under Indian control and has a Muslim majority who wish to be part of Pakistan.
- The part nearest Pakistan is under Pakistani control and has a Hindu majority who wish to be part of India.

- Both areas are separated by a political boundary called **the line of control**. This boundary was agreed by both sides after the United Nations negotiated with the two nations.
- Regular clashes occur, as armies on both sides build up their arms in the face of a political threat. Relations between the two nations are sensitive because:
  - India controls 80 per cent of the Kashmiri population, where there is a majority of Muslims who wish to be part of Pakistan.
  - The headstreams and many tributaries of the Indus river rise in the Indian-controlled part of Kashmir. The Indus is Pakistan's most important river and it depends on it for its water supply and for irrigation. It wishes to gain control of this water source that is vital for its future needs.
  - An increase in Muslim fundamentalism is creating unrest and fighting in the region.
  - Both India and Pakistan have nuclear weapons that could create massive damage and loss of life in a sub-continental region where one-sixth of the world's population lives.

CASE STUDY: GERMANY'S CHANGING BORDERS

- By 1939, before World War Two began, Germany was a rich and powerful state. After World War Two the Allied Forces of Britain and America were determined that Germany would never again be allowed to threaten the peace of Europe.
- So it was divided into two states by the Allies and Russia. This division was part of the Iron Curtain that separated Communist states from the free democratic states of the EU.
  - West Germany was to be a democratic country
  - East Germany was to be a Communist country under Russian control.
  New boundaries were drawn to create these two German states.
- West Germany again became a rich state, where people lived freely in a democratic society. East Germany became poor under a Communist government.
- The East Germans wished they were part of West Germany.

▲ Boundaries of Germany, 1944–1990

- They had little freedom and were not allowed to travel outside their German boundaries. This division of Germany created two new culture groups:
  - East Germans, who were poorer and called Ossies, and
  - West Germans, who were richer and called Wessies.
- Once Germany was reunited in 1990, many East Germans migrated to the western part of Germany in search of work and better living conditions.
- Many West Germans resented these poor migrants, because the reunification of Germany was expensive and a heavy drain on German resources.
- So the once-united German people are now divided again, because of this social division.

SECTION 2:

ELECTIVES

All students must study EITHER

Patterns and Processes in Economic Activities
pages 132–166

OR

Patterns and Processes in the Human Environment
pages 167–202

ELECTIVE 1:
PATTERNS AND PROCESSES IN ECONOMIC ACTIVITIES

ELECTIVE TOPIC 1
PATTERNS OF ECONOMIC DEVELOPMENT

WHAT IS THE MEANING OF ECONOMIC DEVELOPMENT?
Economic development refers to the total quality of life of a population. It includes the standard of its education, medical care and healthy diet. The higher a country's economic development, the better should be the living standard of its people.

MEASUREMENT OF ECONOMIC DEVELOPMENT
- **Gross National Product (GNP)**
- It is measured in US dollars, so that relative comparisons can be made.
- This is called Purchasing Power Parity (PPP).

- **Human Development Index (HDI)**
  This includes three factors as a way of measuring development:
- life expectancy
- GNP per person
- adult literacy rates.

UNEVEN ECONOMIC DEVELOPMENT
- In general, people who live in the northern hemisphere have a high living standard, owing to the development of industry and the gradual urbanisation of the population.
- The majority of people who are in poverty live in the southern hemisphere. Many survive on one dollar a day.

■ Uneven Patterns of Agriculture and Industrial Activities

A country's wealth depends on its levels of:
- agricultural development
- industrial development.

■ The Role of Agriculture

Agriculture is more important in a developing economy than in an industrial economy because:
- There are few alternative employers.
- It supports many people with the basic necessities, even though living standards are at a subsistence level

But it does not improve living standards, as do industry and other occupations, because:
- Only low levels of education are needed to improve output.
- Difficult conditions such as drought or flooding add to the problems of production.
- There is often poor access to markets.
- The price of cash crops is unstable and generally low.

■ The Role of Industry

In contrast to agriculture, the growth of industry improves living standards throughout a population because:
- Once industry is established it encourages a wide range of services.
- Industry and services offer well-paid jobs.
- Transport networks are improved.
- A large home market for products is created by the high wages.
- Industry educates its workforce by introducing new skills.

UNEVEN DEVELOPMENT IN THE EU

Uneven economic development exists among the different EU member states. It also exists within member states.
- There is a core region within the EU where living standards are high. This region includes the Manchester–Milan Axis.
- There are peripheral regions where living standards are lower, for example the West of Ireland or the Massif Central in France.
- There are rich and poor regions within individual countries. In Ireland, Dublin is a rich core region and the West of Ireland or BMW is a poor peripheral region. Similarly the north of Italy is rich and the Mezzogiorno in the south is poor.

ELECTIVE TOPIC 2
CHANGING PATTERNS OF IRISH AND GLOBAL ECONOMIC DEVELOPMENT

Five changes are necessary for a country to improve economically. These are:
1. Change in structure. Primary industries become less important. Industry and services improve.
2. New technologies are introduced.
3. Companies join up to create larger companies, to compete better at home and abroad.
4. Overall quality of life is improved. Better living standards prevail.
5. Volume and value of trade increases.

Fourteen of the twenty poorest countries in the world are located in sub-Saharan Africa, owing to:
*   political instability. Many were former colonies of European powers
*   warfare and civil disturbances
*   droughts, famines and diseases
*   failure to attract industry
*   health-crisis linked to AIDS.
Some countries have shown a measure of improvement. These are called **newly industrialising countries**; they include Hong Kong, South Korea, Singapore and Taiwan.

ECONOMIC DEVELOPMENT AND REGIONAL CHANGE IN IRELAND
Some changes in the economic development of Ireland:

1922–61 Only 10 per cent of people were employed in industry.
A policy of protection was followed. Industries and services located mostly in Dublin.

1961–81 The protection policy was replaced with Free Trade.
This attracted multinationals, and branch plants were established.
Services grew quickly. Dublin and the eastern region benefited most.

1981–91 was a period of economic recession.
Fewer MNCs were attracted to Ireland.
Many traditional industries closed.

1991–2004

During the 'Tiger economy' jobs in industry grew.

A new wave of high-tech industries, such as computer companies and pharmaceuticals, invested in Ireland because of:

- a well-educated workforce
- presence of 3rd-level colleges
- access to good transport services.

Again, all regions benefited from a growth in services. Footloose international service industries, such as call centres, were attracted to Ireland. Again, Dublin benefited the most.

ELECTIVE TOPIC 3
CHANGING PATTERNS OF ECONOMIC DEVELOPMENT IN BELGIUM

The core region of Belgium has changed **from Wallonia in the south to Flanders in the North**. The main industrial centres of Liège and Namur were located in the Sambre–Meuse Valley.

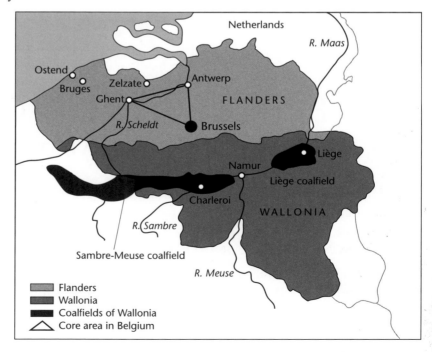

▶ Regions of Belgium showing the Sambre–Meuse coalfield

This change was due to three factors:
- The decline in coal mining and iron and steel production. In the 1950s new oil-burning engines replaced steam engines and so coal was not as much in demand.
- New heavy industries preferred coastal locations, where imports of iron ore could be easily imported.
- The quality coal seams of the Sambre–Meuse valley were exhausted.
- Rising costs of mining made coal uncompetitive as a fuel.
- The growth of service industries created new wealth in the Flanders region.

So now Wallonia is a **depressed region** in the south of Belgium. However, financial support was given by the EU to aid redevelopment. This included:
- modernising its declining steel industries
- retraining workers in new industries
- improving transport routes and the environment
- attracting new industries and services.

WHY DID FLANDERS BECOME THE NEW CORE REGION IN BELGIUM?
The principal reasons were:
1. its central location in Belgium and the EU
2. its location on one of the world's busiest shipping routes, the English Channel and North Sea
3. Antwerp, Belgium's third largest port, is in Flanders
4. its attractive landscape, with the historic towns of Bruges, Ghent, Antwerp and Brussels.
5. a new steelworks at Zelzate on the coast.

ELECTIVE TOPIC 4
COLONIALISM AND DEVELOPMENT

Colonialism led to the exploitation of a large number of countries in the interests of the few colonial powers.

THE REASONS FOR COLONIALISM
- To source raw materials and food produce for industries at home.
- The colonies provided a market for and bought the manufactured goods produced by the colonial powers.
- To gain military control over large areas of the globe.

Example of a colonial power: *Britain*

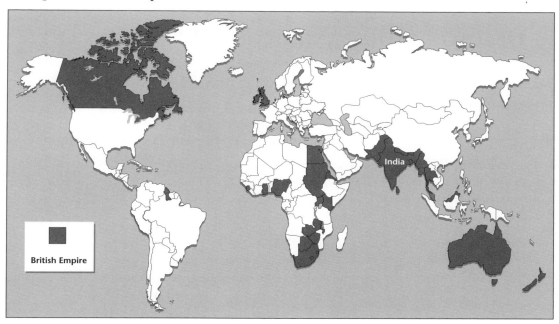

▲ Britain's colonial empire in 1914

■ The Effects of Colonialism

1. Most colonies specialised in producing a few primary products.
2. Native or home industries were neglected.
3. Railways and ports were built in colonies to bring raw materials to the coast for export to the colonial powers.
4. It created a new pattern of world trade called the '**international division of labour**'.

■ Case Study: India – The Changing Economy of a Colony

- Up until 1947 India was a colony of Britain.
- Before colonisation India had many native craft industries and textiles.
- India's native industries collapsed under colonial rule.
- It specialised in producing primary products such as cotton, tea and jute, exported through its major ports of Bombay, Calcutta and Madras.
- It became an independent country in 1947.
- It created its own industries and educated many of its people.
- Today many computer companies and back-office services are located in India.
- Most of its rural regions are very poor and underdeveloped. Its industries are confined mainly to city ports that were developed during colonial times.

ELECTIVE TOPIC 5
DECOLONISATION AND ECONOMIC ADJUSTMENT

How countries coped with the change from being a colony to creating their own trading patterns with other countries.

Many countries gained their independence after World War Two. In order to raise living standards they had to:
1. borrow money to finance development
2. attract new industries
3. create new trade policies.

■ Borrowing
- Many countries borrowed huge sums of money, at high interest rates, from developed countries to build roads, railways, ports and airports.
- Large sums of money that should have been invested in health and education were used to help repay their debts. This made the new states even more dependent on the developed countries.

■ Attracting New Industry
Multinational companies were attracted because they wanted cheap labour to produce basic goods that required few skills. This demand:
- employed many low-skilled people
- increased manufacturing output.

This new trend created a **new international division of labour.**

■ Trade Policies
1. To protect home industries, most developing countries put taxes called 'tariffs' on imported goods.
2. Recently, tariffs were removed and free-trade policies have been adopted. This has encouraged the export of raw materials for which developing countries have a comparative advantage.
3. But today most developing countries still produce mostly primary products.

ELECTIVE TOPIC 6
GLOBAL ISSUES OF JUSTICE AND DEVELOPMENT

Large numbers of people have failed to benefit equally from growing world development.

The developing countries have been unfairly treated in three main areas. They are:
- fair trade
- health services
- gender discrimination.

FAIR TRADE

1. Commodity prices

The prices of primary goods such as coffee and copper have fallen, so poor countries must export more to retain present income. This in turn may glut the market, reducing prices further.

2. Terms of trade

Even though the prices of primary products have fallen, the price of manufactured goods from developed countries has risen. This is a double injustice.

3. Percentage profit

Producers of the raw materials for a manufactured product receive only a tiny proportion of the final price of the product sold to the public.

HEALTH SERVICES

Poor quality of life and low life expectancy are the norm in underdeveloped countries. **AIDS** has become a **disease of the poor.**

GENDER DISCRIMINATION

- In some countries, such as Arab societies, the law discriminates against women. Women do not have equal rights in marriage.
- Many women are not allowed to work outside the home.
- Males have a better chance of going to school than females.
- Infanticide is practised in societies where male heirs are the preferred choice.
- Arranged marriages often force young girls into unwanted relationships.
- Women are seen as a cheap source of labour, especially in the developing world.

ELECTIVE TOPIC 7
GLOBALISATION

The world is now a workplace where decisions made in one part of the world can have major effects on people living in another part.

CAUSES OF GLOBALISATION

- improvements in transport
- improved telecommunications
- more multinational companies
- global banking
- free trade.

ECONOMIC GLOBALISATION

The two key factors in understanding economic globalisation are:

■ **Multinational Companies and their Foreign Investments (FDI)**

Multinational companies invest huge sums of money to set up factories or mines in many countries. This is called foreign direct investment (FDI).

■ **Increased International Trade**

More and more goods and services are being traded worldwide than ever before.

ELECTIVE TOPIC 8
THE GROWTH OF MULTINATIONALS

Some of the largest MNCs have sales that exceed the GNP even of some wealthy countries. Exxon-Mobil, an American MNC, has a business turnover equal to the GDP of a rich country such as Belgium.

Why do Multinational Companies Locate in Different Countries?

■ **Raw Materials**

In the past, colonial powers created colonies mainly as a source of raw materials. Today multinationals source most of their raw materials in less-developed countries for the same reason.

■ **Access to Cheap Labour**

Multinational companies are willing to relocate their branch plants to less-developed countries so as to reduce their production costs. This pattern is called the **'new international division of labour'**.

■ **Access to New, Expanding Markets**

North America, Japan and the EU are the world's largest markets. Multinational companies locate their main production and service plants in each of these regions.

■ **Footloose Locations**

Multinational companies may move part or all of their production from one country to a new location in another where they can manufacture their products more cheaply. This practice gives them great bargaining power when dealing with national governments for grants or tax concessions.

The Product Cycle and Global Assembly Line

1. Initial research and development occurs in the major cities of a developed country, owing to the need for scientists and engineers.
2. Skilled workers and a large market are needed for early product development and sales.
3. As the product becomes simpler, less-skilled workers, lower-cost labour, cheaper land and good transport systems, as well as a large market, become desirable.
4. As the product becomes more basic and easier to assemble, the branch plants are moved to less-developed regions, such as India and South-East Asia, for even cheaper production.

This pattern of location or production can be viewed as a global assembly line. However, not all multinational companies follow the pattern. Dell in Limerick, for example, can manufacture their computers more cheaply than some of their branch plants in South-East Asia, because of more efficient practices in their factory. It takes just five minutes to assemble an entire computer and they manufacture 30,000 units in a single day.

THE INVESTMENT AND LOCATION OF MNCs

There are two main desired locations:
1. major industrialised regions
2. peripheral regions.

MAJOR INDUSTRIALISED REGIONS

About 70 per cent of all MNC investment is located in industrialised regions such as the USA, Japan and Western Europe.

PERIPHERAL REGIONS

Few MNCs initially located factories in developing countries. Since the 1980s this pattern has changed, and by 2000 some 25 per cent of world manufacturing production came from branch plants located in developing countries. The most successful of these developing countries are called **NICs, Newly Industrialising Countries.**

WHAT ARE BRANCH PLANTS?

Branch plants are factories of an MNC that are located in foreign countries, often developing ones. This is done to keep profits high and remain competitive.

THE ADVANTAGES AND DISADVANTAGES OF MULTINATIONAL COMPANIES

■ **The Advantages of Branch Plants**
- They provide work for many people in each factory.
- People learn new skills and new technologies.
- They bring a lot of investment money from abroad.
- They increase exports.
- They create many types of factories and services that help an economy to modernise.

■ **The Disadvantages of Branch Plants**
- Wages may be low, depending on the type of factory.
- Some work is repetitive and boring.

- MNCs may trade only with their branch plants abroad and may not create back-up services locally.
- Much profit is returned to headquarters of the MNC in its home country.
- If branch plants close down, large numbers of people, sometimes thousands, may become unemployed.
- Decision making is generally done elsewhere and the host country has no control over these decisions.

ELECTIVE TOPIC 9
MULTINATIONAL COMPANIES IN THE EUROPEAN UNION AND IRELAND

Many foreign multinational companies have invested in factories in Ireland and in the EU. In addition, many EU multinationals have invested in other regions of the world. These patterns of investment have increased the trade of the EU and helped to make it a richer region.

EUROPEAN MNCs AND THEIR GLOBAL INVESTMENTS

Multinational companies from the EU have been investing in branch plants in other regions for a long time.
- Now the EU is the largest source region of MNC investment.
- More and more MNC funds are used to set up branch plants and back-office services in Newly Industrialising Countries (NICs) in South-East Asia and Latin America.

■ Case Study: Volkswagen (VW)
- Volkswagen is a German MNC.
- It has built many assembly plants and component factories both in developed and developing countries. Trading among these factories has created a global assembly line.
- They include Brazil, Argentina and Mexico in the developing countries of South America. Developed-world regions include the USA, Canada, Shanghai in China and Japan.

■ Foreign MNCs in the EU
- Many foreign MNCs have invested in plants in the EU for access to a rapidly expanding, rich region.

- About 90 per cent of investment comes from the USA, for example Ford, and Japanese companies, such as Toyota.

■ MNCs in Ireland

- MNCs laid the foundations of our industrial development in the 1960s in the Shannon Industrial Estate.
- Production is focused on high-value goods and services such as electronics, pharmaceuticals and internationally traded services.
- New plants were set up for key areas such as research and development.

■ Why did MNCs Come to Ireland in the 1960s?

The main reasons were:
- a large, cheap, but unskilled labour force
- cheap land with plenty of room for expansion of factories
- attractive government grants with tax breaks
- reasonable roads, ports and airports for exporting finished goods and importing raw materials

■ Why do MNCs Invest in Ireland Today?

Because of:
- a large supply of well-educated, skilled, young workforce
- support from high-quality research facilities and universities
- low corporation tax rates for companies
- direct access to the large EU Market.

DELL, A MULTINATIONAL COMPUTER COMPANY

Company name:	Dell Computer Corporation, Ireland
Headquarters:	Austin, Texas in the USA
Main product:	Computer Systems
Global employment:	39,000
World Sales:	US$35.5 billion

Dell produces computers on a global scale. To manufacture and sell efficiently:
- Dell has divided the world outside the Americas into three regions: Europe, Middle East and Africa (EMEA); East Asia; South-East Asia.
- Each region has its own regional headquarters, production and sales network.

■ Dell Ireland

Dell first came to Ireland in 1991.

It has four Irish plants, two in Limerick and two in Bray.

The Bray plants act as sales centre and administration centre for Ireland.

Dell Limerick

Opened its first Irish plant in Limerick in 1991, employing 120 people.

It is the production centre for Europe, the Middle East and Africa.

Its two Limerick plants employ 3,300 people.

30,000 computers per day are assembled and shipped to markets.

Reasons for locating in Limerick

- High-quality University of Limerick, which specialises in Computer Engineering and Manufacturing Systems.
- Suitable industrial estate sites at Raheen and regional Technological Park at Plassey in Limerick.
- Well-developed telecommunication systems.
- Excellent transport systems, such as Shannon International Airport nearby, access to Dublin and Rosslare to export products.
- Government grants and low corporation tax.
- Near other world-class computer companies in Ireland such as Intel, Microsoft and 3com.
- English-speaking country.
- Western European countries such as Germany, Britain and France are the main markets for Dell.

Sources of input for Dell Limerick

- Far East – 52%
- USA – 20%
- Rest of Europe – 11%
- Britain – 7%
- Limerick – 10%.

ELECTIVE TOPIC 10
PATTERNS OF WORLD TRADE

World Trade has hugely increased, especially among the EU, Japan and the USA.

Two factors have been responsible for the increased volume and value of world trade:

- the number and power of MNCs
- improvements in transport and communications.

MERCHANDISE TRADE (TRADE IN PRODUCTS)

Trade is best developed in the developed world.

Regions such as South Asia and sub-Saharan Africa trade little with developed regions because:

- They were colonies until recent times.
- They produce mainly low-value raw materials or basic goods in branch plants of MNCs.
- The prices of most raw materials have declined compared to the price of finished products in developed countries.

The three developed regions of the EU, Japan and the USA, called the Global Triad, control half the value of world merchandise trade. The EU is the most dominant of these three regions.

PATTERNS IN THE LOCATION OF SERVICE INDUSTRIES

Globalisation has increased demands for many services, so international trade in these services is vital for continued global prosperity.

The global pattern of industrial production has increased demand for a wide range of services, such as:

- legal and financial services
- marketing
- research and development
- back-office services.

1. Offshore Financial Centres

Many wealthy people use banks in island states, such as the Cayman Islands or the Isle of Man, to avoid taxation in their own countries. These locations are called offshore financial centres.

2. Geographical Centres of Control

- Major cities in the most developed regions attract the most service industries. These include cities such as London, New York and Tokyo.
- These cities and the global trading triad have used their power to increase their influence over poorer regions of world.

■ **What Factors have Helped to Increase the Trading Power of these Regions?**

- The USA has increased its influence throughout North and South America through NAFTA, the North American Free Trade Area, an organisation that links these regions together.
- The European Union is increasing its trading area through the eastern expansion of the EU and by creating closer ties with Russia.
- Japan is extending its influence throughout South-East Asia, Australia and New Zealand.

ELECTIVE TOPIC 11
THE INTERNATIONAL DIVISION OF LABOUR

■ **What is Meant by the Division of labour?**

All types of jobs or tasks need workers. Then, if workers specialise in certain tasks they become more efficient at those tasks. Consequently, productivity increases and costs fall. This concept is called the division of labour.

■ **What is Meant by the Spatial Division of Labour?**

More profits can be achieved if different areas specialise in certain products or services. Infrastructures and land use are designed to suit this production.

■ **What is Meant by the International Division of Labour?**

If a country specialises in producing certain goods or services for which it has an advantage over other regions, then its productivity increases, its exports increase and its profits increase. This creates money for the country to import goods or services.

■ What is Meant by Comparative Advantage?

Comparative advantage means that a region has some particular advantage for production of specific goods more cheaply than other regions. The advantage may be the presence of **raw materials**, a **specialised labour force** or some other advantage.

WHAT ARE THE FACTORS THAT HAVE LED TO THE DEVELOPMENT OF THE GLOBAL ECONOMY?

There are two key factors:
1. multinational companies and the development and location of their branch plants
2. increased global trade.

■ The Phases in the Development of the International Division of Labour

1. The traditional international division of labour

During the Industrial Revolution the colonies of some industrialising countries, such as England, supplied the raw materials that their factories needed. In turn the colonial powers sold their manufactured goods back to the colonies. In this way the colonial powers became richer.

2. The new international division of labour

The MNCs of core regions established branch plants in peripheral and developing economies, so as to concentrate home production on higher-level goods and services.

3. The newer international division of labour

From the 1990s onwards a range of back-office services, for which low wages are vital, have been relocated from core countries to peripheral and developing countries. The positive response from the developed countries to this relocation is to specialise increasingly in high-tech industries and higher-order services.

4. The most recent international division of labour

MNCs are subcontracting an increasing number of key functions from highly paid locations in core countries to peripheral regions where work forces are becoming well educated. These functions include high-value and skill-demanding jobs such as research, computer-chip design and financial services.

A most important factor in this new phase is that work is coming from developed regions to educated people in developing peripheral countries, rather than well-educated workers having to migrate for work to core countries and causing a brain drain at home.

■ **Why do MNCs Locate their Branch Plants and Back-office Services in Peripheral Regions?**

- Labour costs are low.
- There are large workforces due to high birth rates.
- There is little union input, so workers can be exploited.
- Many developing countries have increasingly tried to educate their people so as to attract higher-value jobs.

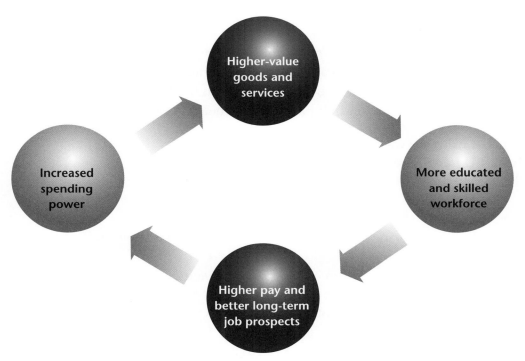

▲ Advantages of the phases of the international division of labour for developing countries

■ **Case Study: Nike, a Footwear and Clothing Company**

Headquarters in Bevertown, Oregon, USA.

16,000 directly employed and 500,000 indirectly employed.

First: Original locations of production plants have closed because labour costs became too high.

Second: 30 newer locations opened in South-East Asia, for example South Korea and Taiwan. Production costs here increased, so the company moved to less expensive countries.

Third: Relocated to even cheaper locations, such as Indonesia, Thailand and China.

International group protests embarrassed Nike into joining the **Fair Labour Association.** This association promotes the ideal of a basic minimum wage to meet daily needs at least.

ELECTIVE TOPIC 12
TRADING PATTERNS IN THE EUROPEAN UNION

The EU accounts for 40 per cent of all trade in goods and services worldwide. Because a number of European countries were colonial powers, many of their former colonies are still major trading partners with the European Union. So 40 per cent of the growing global trading patterns still focus on Western Europe.

Patterns of EU trade can be divided into two types:
- **Intra-EU trade** – trade between member states.
- **Extra-EU trade** – EU trade with the rest of the world.

INTRA-EU TRADE

Since World War Two, the growth of trade between member states has grown rapidly because of:
- the Treaty of Rome creating free trade between member states
- enlargements of the EU
- well-developed rail, road, pipeline and air routes
- the large size and wealth of the EU
- the creation of the Single European Market in 1993
- eastern expansion of the EU.

EXTRA-EU TRADE

1. **Colonial links** have created a web of trade routes with the EU.
2. **Japan and the USA** are the EU's most important trading partners, creating a Global Trading Triad.
3. The **Lome Convention** in 1963 created trade links between ACP countries. These are countries in Africa, the Caribbean and the Pacific. Most of them were former colonies of EU countries.
4 There is increased trade between the EU and **MNC branch plants** in Asia and Latin America.

ELECTIVE TOPIC 13
IRELAND'S TRADING PATTERNS WITHIN THE EU

- Before the 1960s Ireland used a tariff or tax on imported goods to protect home industry from competition.
- Today it is part of a tax-free open market in which all EU countries export to each other without any additional tax.

1. CHANGES IN THE VALUE OF IRISH TRADE

Before the 1960s the value of Irish trade was small. Ireland's imports were more expensive than the value of its exports because:
- of the high costs of imported oil and coal
- MNC branch plants used cheap labour to manufacture low-value goods
- MNCs mined minerals for export rather than for processing.

A huge rise in the value of Irish exports in the 1990s happened because:
- There was a change from branch plants producing cheap goods to making higher-value goods and services.
- This increased taxes, so the government could invest more in motorways, urban transport links and urban bypasses.
- A rapid rise in trade value created the **Celtic Tiger** economy.

2. IRELAND'S CHANGING PATTERN OF TRADE

- Until the 1980s Ireland was dependent on Britain for over 50 per cent of the value of its exports.
- Today it makes up 36 per cent of exports by value. So we are less dependent on Britain, but it is still very important to us.
- MNCs have created a global trading pattern with Ireland through their branch plants.

3. CHANGES IN THE MAKE-UP OF IRELAND'S EXPORT TRADE

- Until the 1970s food and live animals formed the largest part of our exports.
- EU membership and MNC operations, such as electronics and chemicals, increased industrial goods and services.
- Agricultural products now make up only 6 per cent of exports.
- Almost all MNC products are for export, and the EU market is hugely important to them.

ELECTIVE TOPIC 14
THE COMMON AGRICULTURAL POLICY AND ITS IMPACT ON IRELAND

The Common Agricultural Policy (called the CAP) was introduced in 1962. Its aims were:
- to increase production and productivity
- to provide a fair living standard for all its farmers.

WHAT POLICIES WERE NEEDED TO ACHIEVE THESE AIMS?

It was necessary to:
- introduce a common tariff (tax) on all imports from outside the EU, to protect farmers from cheaper imports.
- establish a Guarantee Fund and a Guidance Fund to finance the CAP.

WHAT DID THE GUARANTEE FUND DO?

- It bought any surplus farm produce within the EU to maintain high prices that were fixed each year.
- It subsidised exports throughout the world and so reduced stored surpluses. These stored surpluses were called 'intervention'.

■ What did the Guidance Fund do?
- This fund provided money to modernise farm buildings and machinery and to organise farms into single, larger farm units rather than scattered smaller ones.

Later changes to the CAP

Changes introduced later were:
1. reduced prices guaranteed to farmers
2. diversified farm activities, so as to create more income for farmers, e.g. tourism, cheese-making
3. making farmers more competitive
4. creation of more direct income support for small farmers
5. protection of the environment.

IMPACT OF THE CAP ON IRELAND

Positive effects

- Farms were modernised and productivity was increased.
- Subsidies changed some farmers from dairying to sheep rearing, especially in upland regions and the West.
- There was a major increase in the value of Irish exports.
- Farm incomes increased as their output increased.
- The number of small farms was reduced.
- There was increased income for small farmers by means of subsidies.
- Farm sizes increased.
- Farms became more specialised.

Negative effects

- Inequality between small and large farmers was increased.
- The number of farmers was reduced.
- There was increased migration from rural areas.

THE EFFECTS OF CAP ON THE ENVIRONMENT

Negative effects

- Increased use of fertilisers led to soil and water pollution.
- Hedgerows and stone walls were removed to increase field size.
- Habitat for wildlife such as plants and animals was reduced.
- Overstocking of the land led to overgrazing and soil erosion in hilly areas.
- Old, traditional farm buildings were replaced by modern structures

Positive Effects

- Reps programme was introduced recently to protect the environment.

ELECTIVE TOPIC 15
THE COMMON FISHERIES POLICY AND ITS IMPACT ON IRELAND

THE NATURAL ADVANTAGES OF IRELAND'S OFFSHORE SHELF FOR FISHING

- Shallow waters allow light to reach the seabed to create food supply in the form of plankton.
- A warm North Atlantic Drift attracts a wide variety of fish species.
- Calm and sheltered regions encourage fish spawning and mariculture.
- Some of the richest fishing grounds of the EU occur in Irish waters.

■ **How has the Common Fisheries Policy (CFP) Affected the Irish Fishing Industry?**

Negative effects

- It has restricted the development of the fishing industry in the following ways:
- It has exposed Irish waters to major fishing fleets of countries such as Spain.
- With 11 per cent of EU waters it has only 5 per cent of the Total Allowable Catch (TAC).
- Over-fishing has almost wiped out many fish species, such as cod and herring.
- The Irish fishing fleet is made up of a small number of large vessels that catch most of the fish landed.
- The larger vessels have led to further over-fishing.
- Fishing is concentrated out of a small number of major ports.

Positive effects

- The total value of fish landings has increased significantly.
- A small number of fishermen have become very wealthy, and are called mackerel millionaires.

THE COMMON REGIONAL POLICY OF THE EU AND IRELAND

- When Ireland joined the EU it was the poorest of the member states.
- Since then, large transfers of Structural Funds have been responsible for Ireland's development.

■ **Structural Funds of the Common Region**

- **ERDF:** the **European Regional and Development Fund**, to aid industrial development and upgrade roads.
- **ESF:** the **European Social Fund**, to train/retrain workers who become unemployed in problem regions.
- **FIFG:** the **Financial Instrument of Fisheries Guidance**, helps the fishing industry and fishing regions.
- **Guidance Section of the Agricultural Fund**: to improve farm structures.

■ The Reformed CRP of 1989–99

This helped Ireland in the following ways:

- Ireland was designated an Objective 1 region for guaranteed structural funds. Objective 1 regions are least-developed regions and must have a GDP per person of less than 75 per cent of the EU average.

- Ireland must submit National Development Plans to receive funds for:
  1. modernising high-tech industry
  2. improving transport and communications
  3. increasing labour skills for future challenges of change.

■ Changes to Ireland's Objective 1 Status

- Structural funds were reduced because of Ireland's increased wealth.
- The BMW region is the only Objective 1 region from 2005 onwards.

■ The ESP, the European Social Fund

Through its National Development Plans the ESF hopes to:
- reduce employment through training schemes
- provide affordable housing
- integrate minorities
- create gender equality in the workplace
- provide community support schemes for people in disadvantaged urban areas.

ELECTIVE TOPIC 16
RENEWABLE AND NON-RENEWABLE RESOURCES

Renewable resources are those such as water that, when used wisely, can be used over and over again.
Non-renewable resources are those such as oil and natural gas that cannot be renewed.

TRENDS IN ENERGY RESOURCES IN THE EU

- Oil and natural gas have replaced coal as the main source of energy.
- Nuclear power has increased rapidly for those countries that lack other energy resources.
- Owing to increasing affluence, almost half of EU energy demands must be met by imports.
- Most EU countries must import energy supplies to meet their needs.

■ Ireland's Energy Resources

- Bord na Mona, which provides low-grade fossil peat fuel for domestic use and for three peat-fired power stations
- six hydroelectric power stations

- one coal-fired power station
- three oil-burning power stations
- three natural-gas power stations
- three gas fields: Mayo, Seven Heads and Kinsale
- new wind turbine locations
- only 15 per cent of energy needs met by home resources.

THE ENVIRONMENTAL IMPACT OF BURNING FOSSIL FUELS

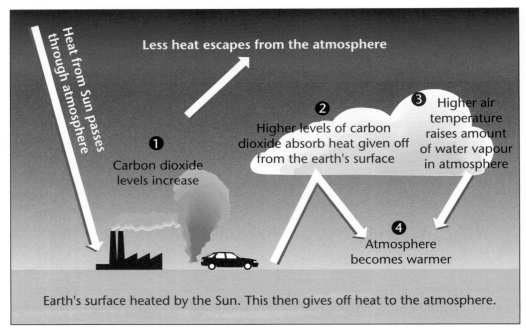

▲ Flow chart of carbon dioxide in the atmosphere

■ Acid Rain

The burning of fossil fuels releases nitrous and sulphurous gases into the air. These gases dissolve in rainwater acids to form acids that fall to the ground and have the following negative effects:
- Forests are damaged, because:
  - tree growth is stunted
  - leaves are discoloured and fall early
  - bark splits and is affected by cold weather.
- Acid levels in rivers and lakes rise and kill aquatic life.
- Essential nutrients are leached from soils.

- Toxic minerals enter rivers and lakes.
- Buildings are damaged.
- Respiratory diseases such as asthma increase.

■ Greenhouse Gases

Greenhouse gases, especially carbon dioxide, are released into the air. These are leading to:
- an overheating of the global atmosphere, quickly changing climates and melting of ice caps. This process is called global warming
- smog and smoke pollution in urban regions. This has led to the use of smokeless fuels in urban regions such as Dublin and Limerick.

■ The Rise in Carbon Dioxide in the Atmosphere

This is caused by:
a. industrial growth and higher living standards in the developed world
b the spread of industrial development into less-developed world regions
c. higher personal demands for energy
d. rapid growth of the world population.

■ Some Consequences of Global Warming

Climate changes may lead to:
- droughts and famines in sub-Saharan Africa
- sea-level rises and flooding of low-lying coastal areas
- melting of ice caps
- increases in tropical diseases
- higher cancer risks.

■ Some Solutions to Global Warming

a. Worldwide agreements such as the Kyoto Protocol.
b Reduce the burning of fossil fuels.
c. Promote the use of clean alternatives, such as wind energy.
d. Reduce the process of deforestation.

■ Ireland and the Kyoto Protocol

a. Ireland must reduce its 1990 levels of CO emissions by 13 per cent by 2012.
b. Reduce the burning of coal at Moneypoint power station.
c. Increase the area of forestry to absorb carbon dioxide and release oxygen.
d. Increase the use of alternative energy sources, such as wind power.
e. Introduce a 'carbon tax'.

ELECTIVE TOPIC 17
RENEWABLE ENERGY AND THE ENVIRONMENT

■ The Difficulties of Increasing the Use of Hydroelectric Power Supply

a. Most of the best sites are already developed.
b. High costs of dam construction and reservoirs.
c. Environmental problems such as wildlife habitats and fish migration are badly affected.
d. Local community disruption. People may have to be relocated and rehoused.
e. Loss of land due to flooding behind the dam.
f. Visual pollution from dam structure.

■ Wind Energy in Ireland

- Wind energy is the most preferred clean energy alternative in Ireland at present.
- Seven per cent of Ireland's energy is generated by wind in 2005.
- Two hundred wind turbines are to be built on the Arklow Bank, offshore, and will generate enough energy to meet the needs of 500,000 people.

The most suitable sites for wind turbines

a. large, uninhabited or low population density regions such as hill tops
b. exposed coastal or inland sites with constant strong winds
c. offshore shallow water banks.

Some disadvantages of wind farm development in Ireland

- noise caused by rotating turbine blades
- visual impact of turbines
- mass movement of surface material, such as peat, when disturbed for development
- damage to homes and life due to mass movement.

ENVIRONMENTAL POLLUTION
Pollution can occur at local, national, international and global levels.

■ Pollution at Local and National Levels

Waste disposal

As Ireland's prosperity has increased, so have its levels of pollution. There is an urgent need to find new ways to dispose of waste because:

- most existing waste-disposal sites are already near capacity levels
- most communities are opposed to new landfill sites nearby.

■ Ways to Dispose of Waste

Incineration

Advantages of incineration

– takes up little space
– capable of huge volume of waste disposal
– generates heat for additional energy supply
– burning at high temperatures creates only limited pollution.

Disadvantages of incineration

– increases air pollution
– releases dioxins into the air that may cause cancer
– toxic ash must be disposed of.

Six major incinerators, one for each of six regions, are planned to deal with waste in Ireland.

One at Poolbeg in Dublin could:

1. treat 25 per cent of the city's waste
2. generate energy for 35,000 homes.

Most people are afraid that incinerators would damage the local environment and people's health.

Recycling

- Ireland's volume of waste is unsustainable for the future.
- Irish people need to become more committed to recycling.
- Irish people are the least committed within the EU to recycling.
- Government aims to recycle about 45 per cent of waste by 2010.

■ Radioactive Pollution

Radioactive particles are dangerous for at least 30,000 years.

The three main sources of radioactive pollution

- discharge from nuclear power plants into the air and sea
- disposal of radioactive water created within plants
- decommissioning of nuclear plants. Buildings and materials remain dangerous for thousands of years.

Other causes for concern

- accidents caused by human error leading to discharge
- normal leakages as a consequence of daily routines and wear and tear.

■ The Effects of the Chernobyl Nuclear Accident

- The surrounding region is the most radioactive in the world, so it is permanently damaged.

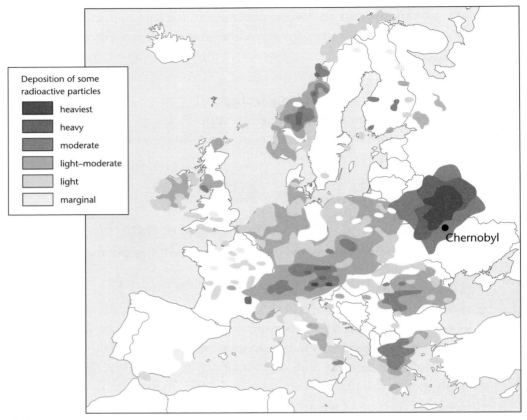

Deposition of some radioactive particles

- heaviest
- heavy
- moderate
- light–moderate
- light
- marginal

Chernobyl

▲ The distribution of radioactive fallout from the Chernobyl accident

- Twenty thousand towns and villages remain uninhabited.
- Large areas of forest and farmland are radioactive and cannot be used.
- Seven million people have been affected.
- Cancer cases and birth defects are abnormally high.
- Regions far from Chernobyl, such as Donegal, were affected by fallout.

Activity

- Name the country in which Chernobyl is located.
- What levels of contamination did Ireland receive from the Chernobyl fallout?
- Explain why Ireland and Norway received higher radioactive fallout than Britain or Sweden. (Hint: cyclone air movement)

■ Radioactive Pollution from Sellafield

What does the Sellafield plant do?

1. It reprocesses used nuclear waste.
2. It stores used nuclear waste from Britain and abroad, for example from Japan, for reprocessing.

3. It stores plutonium and enriched uranium for sale to nuclear plants.
4. It receives by sea used radioactive uranium that travels through the Irish Sea.
5. It discharges waste radioactive water into the Irish Sea, making it the most radioactive sea in the world.

ELECTIVE TOPIC 18
SUSTAINABLE ECONOMIC DEVELOPMENT AND ENVIRONMENTAL ISSUES

Sustainable economic development is a long-term plan that is vital for reducing the impact of people's activity on a region's natural resources.

■ **Environmental Impact Studies in Ireland**
- These form a vital and integral part of national and county development plans.
- They are carried out by independent researchers.
- They assess and report on the state of the environment.
- They look at the costs and benefits of any new project.
- They estimate its possible impact on the environment.

■ **The Role of the EPA (Environment Protection Agency)**
1. To promote and implement the highest practical standards of environmental protection and management for sustainable and balanced development.
2. To license and control all large-scale activities that could impact on the environment.
3. To ensure that all waste-disposal sites apply for a pollution-control licence to operate.
4. To record and monitor all industrial emissions from individual factories.

IRISH FISH STOCKS AND SUSTAINABLE DEVELOPMENT
■ **Why have Irish Fish Stocks Decreased?**
- Irish fish landings have increased four-fold over a thirty-year period.
- There has been an increase in the size and efficiency of fishing vessels.
- Unrestricted access has been granted to Irish waters by large EU fishing fleets.
- New research and monitoring of fish has provided details of shoal movements at various times of the year.

■ **How can Fish Stocks be Sustained?**

- Allow spawning stock to reproduce at an effective level.
- Allow young fish to reach maturity so they can breed and multiply.
- Increase mesh sizes on nets.
- Make the use of all monofilament nets illegal.
- Reduce net sizes.
- Reduce the existing total allowable catch and national quotas for each fish species in EU waters.
- Create exclusion zones where fish can spawn.
- Provide protection vessels to control and enforce conservation measures.

MINING AND ENVIRONMENTAL IMPACT

Past mining operations created unsustainable development because:
- High waste tip-heaps that encircled settling ponds were owned by the mines. These created visual pollution.
- Water used for treating mineral ores was released into nearby rivers.
- Exposed mining pits filled with polluted water once the mines closed.
- A high dust content was created in the air close to the mines.
- Local mining villages with a high unemployment rate depended on the mine for unskilled work – for example Silvermines village in Co. Tipperary – but created a spoiled landscape that repelled new industry to the area.

■ **Case Study: Tara Mines in Co. Meath**

Began operation in 1977 and is the largest zinc mine in the EU. Zinc ore is extracted and processed to concentrate ore (almost pure zinc).

Severe planning restrictions were applied to planning permission
The reasons for this were:
- It was centred in a fertile farming region for fattening cattle.
- It is close to the Blackwater, a major fishing river.
- it is near large urban regions such as Navan town.

Planning restrictions included:
- new tree plantations to screen the development from public view and reduce dust distribution
- noise and air pollution to be closely monitored
- water used in the mine to be purified before being released into the Blackwater
- large quantities of mining waste to be returned underground or contained in environment-friendly settling ponds
- water from the tailings pond to be recycled in the mine.

ELECTIVE TOPIC 19
CONFLICTS OF INTEREST DUE TO RESOURCE DEVELOPMENT

FISH FARMING IN IRELAND

■ **Reasons for the Growth of Fish Farming**

- The introduction of quotas on fishing fleets led to a demand for increased fish supply.
- Higher living standards led to a need for a balanced diet that includes fish products.
- Ideal conditions exist for fish farming, including:
  - pollution-free waters off the west and south coasts
  - numerous sheltered bays and estuaries
  - regular tides that help flush out toxic waste from fish populations.

■ **Economic Advantages of Fish Farming**

1. It creates employment for many coastal communities in the west of Ireland. Over 1,800 people are directly employed in fish farming.
2. Spin-off industries such as fish-cage manufacture, preparation of fish feed and supply and fish processing add to employment numbers.
3. Generous government incentives are available for setting up fish farms.
4. There is local expertise in the fishing tradition.
5. Large quantities of fish can be reared in numerous coastal bays.

■ **Environmental Objections to Fish Farming**

1. There were few objections initially, as environmental effects were not fully understood.
2. Water quality declines owing to the addition of chemicals for flesh colour and prevention of fish diseases.
3. Toxic waste from excess chemicals, fish excrement and dead fish builds up directly under the fish cages.
4. Disease spreads to local natural fish populations: for example, the spread of increased fish lice to sea trout that has almost wiped out this species in Irish waters.
5. Scenic locations can be visually polluted by fish cages.
6. Interbreeding of farmed salmon with wild salmon can interfere with the salmons' ability to survive as a species.

Case Study: A fish kill in Inver Bay in Co. Donegal

- In 2002 about 50,000 farmed salmon died in Inver Bay.
- In 2003 about half a million farmed salmon died at a fish farm.
- These dead fish totalled 2,200 tonnes of dead matter.
- No definite cause of this fish kill was found.
- This lack of information or cause poses a threat to all wild fish that enter Irish waters or live nearby.

Conflict of interest

1. Many jobs depend on continued fish farming.
2. Other forms of fishing, such as inland salmon and sea-trout fishing, from which a greater income is gained, may be at risk.
3. Tourism is affected by interference in the natural landscape.

Development of Irish Bogland

Conflict arises over the way Ireland's bogs should be used.

■ Economic Advantages of Bogland Development

1. It creates employment in Ireland's Midland region, where few industrial jobs are available.
2. Peat is a major source of cheap fuel for many disadvantaged communities in the Midlands and West.
3. Energy is generated from peat-fired power stations, reducing the amount of imported fuels needed.
4. Many families depend on turf supplies for winter heating.

■ Environmental Advantages of Unspoilt Bogs

1. They support a wide range of rare plants and animals.
2. They act as bird sanctuaries at night for many migratory birds, such as wild duck.
3. Many of Ireland's archaeological sites are preserved under their peat cover.
4. They are natural landscapes that attract many specialist tourists to study the flora and fauna.

GLOBAL ENVIRONMENTAL CONCERNS

■ The Economic Advantages of Deforestation

- Powerful logging companies have much profit to gain by this activity.
- Developing countries such as Brazil increase their revenue for other development by the export of forest products.
- Large ranches are developed on cleared forest soil for the grazing of cattle for low-cost meat for developed-world markets.

- New farms can be supplied to poor, landless peasants.
- Consumers in the developed world provide a major market for tropical hardwood products.

■ Economic and Environmental Disadvantages of Large-scale Deforestation

1. Large-scale loss of soil is caused by exposure to torrential tropical daily downpours. Upper soil layers are washed away.
2. The absence of large quantities of tree litter leads to a quick loss of natural humus that enriched the soil through decomposing leaves.
3. The felling of trees for agriculture leads to slash-and-burn activities, fast loss of soil fertility and finally abandonment of farms.
4. There is a loss of vast quantities of plant, animal and human environments that were in balance with nature.
5. There is also a loss of pharmaceutical cures from plant life that could lead to medical cures for cancer and other diseases.
6. Contact with modern society threatens the survival of primitive forest tribes who live in the rain forests.
7. Tropical forests supply a large amount of oxygen to the atmosphere. They also absorb massive amounts of carbon dioxide from the atmosphere, so reducing global warming.
8. Forests add a large quantity of water vapour to the atmosphere through the process of transpiration. An imbalance due to loss of forest cover could dramatically affect our human and wildlife habitat and even threaten our existence.

■ Policies for Sustainable Development

1. Promote the resources of the rainforests as a source of employment, health and wealth by the harvesting of selective plants for medicinal needs and food supply.
2. Establish national parks so the natural landscape is preserved and for the development of high-income tourism.
3. Give financial incentives to countries that preserve their natural forests by offsetting debt versus preservation.

DESERTIFICATION

■ Causes and Consequences of Desertification

- Unreliable rainfall has led to repeated drought.
- Overcropping and overgrazing of land led to a clearance of vegetation.
- The dry, light soil was blown away, leaving only coarse-grained sand and gravels behind.

- Failure of the land to provide sufficient harvests and grazing led to famine. Millions died during the subsequent famines.
- Many people were forced to migrate to regions to the south. This led to overpopulation in these regions.
- The supply of wood to urban populations led to the cutting down of scrub vegetation. This led to increased wind speeds, which in turn led to soil erosion.
- The demand for primary commodities such as groundnuts and cotton encourages large-scale tillage farming in marginal regions.
- Governments support cash cropping to gain income from exports.
- Falling world prices for cash crops leads to increased tillage.
- Soil fertility declines as cash crops are grown year after year in the same areas.

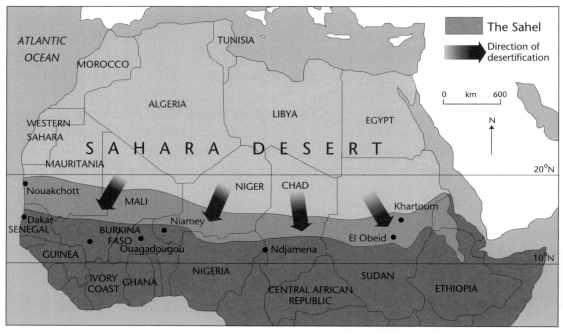

▲ North Africa, showing the Sahel and the direction of desertification

■ Solutions to Desertification

- Reduce dependence on cash crops for exports.
- Increase afforestation, i.e. the planting of large areas of pine forests to create shelter belts and reduce wind speed to prevent soil erosion.
- Fence off large areas for a number of years from grazing animals, to allow vegetation to re-establish itself.
- Provide irrigation projects to help increase food supplies for traditional farmers.

ELECTIVE 2:
PATTERNS AND PROCESSES IN THE HUMAN ENVIRONMENT

ELECTIVE TOPIC 20
WORLD POPULATION DISTRIBUTION AND DENSITY

- The world's surface is very unevenly populated.
- About 80 per cent of the world's population occupies about 10 per cent of the world.

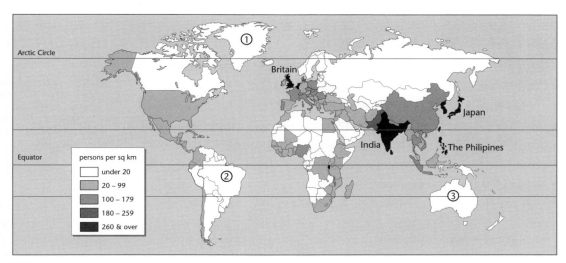

▲ World population-density distribution

Activity
Why have regions 1–3 low population densities? In your answer refer to
- climates
- vegetation.

■ **Four Most Populated Regions of the World:**

1. Western and central Europe
2. Eastern USA and south-eastern Canada
3. The Indian subcontinent, including Pakistan, India, Sri Lanka and Bangladesh
4. East and South-East Asia, including China, Korea, Japan, Malaysia, the Philippines and Indonesia.

■ **The Least Populated Regions of the World:**

1. The cold tundra of Northern Canada, Greenland, Siberia and Antarctica
2. Mountainous lands such as the Rocky Mountains in the USA and Canada, and the Himalayas
3. The plateau lands of Tibet and Central Asia
4. Hot desert regions of Australia, the Sahara and the Arabian peninsula, and the deserts of Iran and Pakistan
5. The equatorial rain forests of the Congo basin in Africa and the Amazon in South America. The Amazon Basin includes much of Brazil and parts of Peru, Ecuador, Colombia and Venezuela.

Physiologic density refers to the ratio of people in a country per unit of area of agriculturally productive land.

THE EFFECTS OF URBANISATION ON WORLD POPULATION DISTRIBUTION

- In 1950, 29 per cent of the world's population lived in urban areas.
- In 1990, 43 per cent lived in towns and cities.
- In 2000, 50 per cent lived in towns and cities.
- By 2030, more than 60 per cent of the world's population will live in urban areas.
- In 1960 there were two cities with a population in excess of 10 million people.
- In 2004 there are seventeen such cities.
- In 2015 there will be twenty-six.
- All mega-cities lie within 500 kilometres of a coastline.

The highest population densities occur in India, Korea, Japan, the Philippines and Britain.

SOME EFFECTS OF MIGRATION ON POPULATION DISTRIBUTION AND DENSITY

1. Millions of Europeans migrated to the United States and Canada in the eighteenth, nineteenth and twentieth centuries. This created a high density of population in eastern USA and south-eastern Canada.

2. Over 6 million people from southern Italy have migrated to northern Italy over the past 50 years. This has had two effects:
   - It has increased the population of northern Italy.
   - It has reduced the population of southern Italy.
3. Millions of people have migrated from Ireland since famine times. This has reduced the overall population and density especially in the west, north-west and midlands regions.
4. Spanish and Portuguese colonisation of Latin America has led to
   - a large density of people of European ancestry in this region
   - a low density of Native Americans in this region, due to the spread to this new world of European diseases that wiped out native populations.

ELECTIVE TOPIC 21
PATTERNS IN THE GROWTH OF POPULATION

■ Why were Population Numbers Low for a Long Time?

- Initial populations were small, so this limited the rate of population growth.
- Agriculture was discovered some 12,000 years ago in the Middle East and then spread throughout Europe and Asia. However, other world regions, such as sub-Saharan Africa, the Americas or Australia, did not discover agriculture.
- Famines, diseases and war between neighbouring groups and by imperial powers kept population growth rate to a minimum.
- Before transport improvements, food supply and types of food available were limited to the regions in which they were grown. Not many people were well off so the available food supply limited the number of people in families.
- Persecution, local laws and the power of colonists' laws all contributed to the deaths of many people.

■ Why did World Population Grow Rapidly from 1750 Onwards?

1. New farming methods, such as selective breeding and creation of individual farm units, prevented the spread of animal diseases.
2. Improved technology, such as seed drill machines, created increased output from farm units.

3. The invention of the steam engine led to increased employment, affordable and better housing in urban regions and a corresponding rise in population.
4. Improved hygiene and medical knowledge were gained.
5. There was increased land supply in the New World.

■ **Why did World Population Grow Rapidly in the Twentieth Century?**

- There were great improvements in medical care, such as antibiotics and the control of many diseases, for example tuberculosis.
- New high-yielding seed varieties increased food supplies.
- Increased clean water supplies and better sewage-disposal systems were developed, leading to control of infectious diseases.
- Lower death rates and increased life expectancy led to rapid population growth rates.

■ **Why do Population Growth Rates Vary between Regions of the World?**

- Growth rates for developing countries such as India are high because they are in the early, expanding phase of population growth.
- Birth rates are very high: 2.8 per cent growth rate for Pakistan, which causes a population to double in 25 years.
- Growth rates for developed regions such as the EU are stable. Growth rates for individual countries are declining, as in Germany. They are in the final or senile stage of the population cycle.

ELECTIVE TOPIC 22
CHANGING POPULATION CHARACTERISTICS

ANALYSES OF AGE–SEX PYRAMIDS

These are useful for the following reasons:

■ **Birth Rates indicate the potential:**

- population of a country for many years ahead
- school-going population and the number of teachers required for the future
- paediatric care needed in hospitals and the number of doctors.

■ **Death Rates indicate:**

- the standard of medical care that reflects the wealth of a country
- the number of pensioners or elderly dependants
- the number of nurses, doctors and nursing homes needed for the future.

■ **Age–sex Pyramids indicate**

- migration patterns reflecting movement into and out of the country
- dependent age groups and size of working population
- balance between males and females.

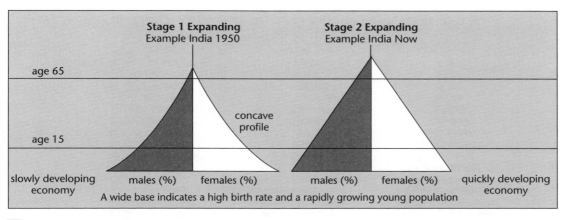

▲ Age pyramids showing population of India in 1950 and now

1 Few old people. Death rates high in all age groups.
2 No social welfare system.
3 High birth rates and high death rates indicate a poor, undeveloped country. Few industries.
4 Longer life expectancy. Larger number reach over 65 years.
5 Fall in death rates causes growth in middle age groups. Fast-growing population.
6 Still high birth rates.

Dependency ratio

- This ratio is the number of children under 15 and people over 65, relative to the working age group.
- In developed countries the young dependency group rises as school and college-going ages rise.
- The greater the number of the dependent age group relative to the workers, the larger the number of people being supported by a smaller number of workers.
- As populations grow older, so the cost of caring for the elderly in that country rises. This increases taxes on the workers in order to provide for them.

ELECTIVE TOPIC 23
PATTERNS OF POPULATION CHANGE IN IRELAND

The Republic of Ireland's population declined from 6.5 million in 1841 to under 2.8 million in 1961.

STAGES IN IRELAND'S POPULATION-GROWTH PATTERN

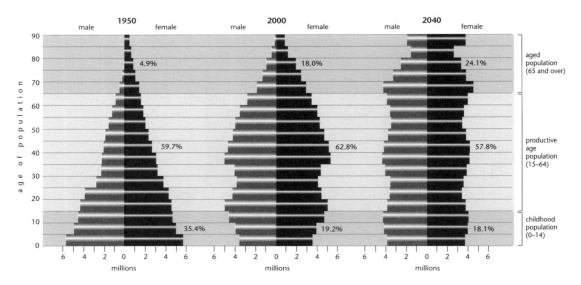

▲ Graph of stages in Ireland's population-growth patterns, 1950–2040

1. Famine and emigration led to a rapid fall in Ireland's population in the 1800s.
2. Improved food supply and medicines slowed population fall. Many continued to emigrate owing to a lack of jobs.
3. A severe depression in the 1950s forced over 40,000 people to emigrate annually.
4. A new economic policy and attraction of MNCs led to increased employment that halted emigration.
5. An economic depression in the 1980s forced many to emigrate, especially the young.
6. Celtic Tiger economy attracted many foreign nationals and returning Irish from abroad. This led to a rapid rise in population.

■ Age-structure Changes

There are clear changes in the age structure of Ireland's population since 1961.
There was a high dependence ratio in 1961 in Ireland.

- A large number of people in the 65+ category had to be supported by few workers.
- Huge numbers of young people emigrated to the United States, Australia and Britain once they reached working age. This left a great shortage of people of working age. The country suffered from a 'brain drain'.
- Ireland had a very high birth rate at this time. Few went to third-level colleges.

There is a lower dependence ratio in 2002 in Ireland.

- There are more people in the 65+ category. Medical care is improving, resulting in fewer deaths.
- There is a very large workforce, with no emigration and some immigration.
- Ireland has become a multi-ethnic society for the first time.
- Many mothers work outside the home.
- The birth rate is low and stable, but still 2 per cent greater than in Japan, indicating an industrialising economy in Ireland.

Life expectancy for both males and females in Ireland remains low compared with many other EU countries.

■ Ireland's Dependency Ratio

- This has fallen since the 1960s because:
- There is now a large bulge in the age–sex pyramid for 2002, indicating a large and growing workforce.
- Immigration of young working adults has added many to the workforce.
- Many students work part-time, earning income that helps support them in school and college.
- There is a dramatic rise in the number of married women who work outside the home.
- Increased personal pension allowances will help to reduce expenditure on the ageing population over the next few decades.

ELECTIVE TOPIC 24
THE CAUSES AND EFFECTS OF OVERPOPULATION

■ **What does Optimum Population Mean?**

Optimum population is the number of people working with all the available resources of that area, who will produce the highest standard of living and quality of life available to them.

■ **What does Overpopulation Mean?**

Overpopulation occurs when there are too many people in an area for the resources of that area to maintain an adequate standard of living.

1. THE IMPACT OF RESOURCE DEVELOPMENT

■ **Case Study 1: Causes of Overpopulation on the Aral Sea**

- There was over-development of water supplies for irrigation.
- Numerous canals were built to divert water from two rivers, the Amu and Syr, to provide water for irrigation of cotton, rice and melons

Consequences

- Only tiny amounts of fresh water now reach the Aral Sea.
- The ecosystem of a fresh-water sea was destroyed and changed to a salt-water sea.
- Fish life, delta farmland, forestry and swampland were destroyed.
- Coastal fishing communities were destroyed, jobs lost, and ships stranded on a dry seabed.
- Thousands of people migrated from the area in search of work. Many urban areas turned into ghost towns.

■ **Case Study 2: Causes of Overpopulation in the Sahel**

- Unreliable rainfall and shortage of rainfall led to increased grazing on limited grassland areas.
- Grass was overgrazed, leaving soil exposed to wind erosion.
- Falling prices of cash crops led to increased tillage in lands with unreliable rainfall.
- Demand for wood led to trees and bushes being cut down for firewood.

Consequences

- Unreliable rainfall led to drought and famine.
- The felling of trees and bushes led to increased wind speeds that aid erosion.

- Increased tillage exposed the soil to wind erosion.
- Millions died of hunger, disease and drought.
- Cash crops failed also, leaving the governments of affected countries in increased debt.

2. UNDERDEVELOPMENT OF RESOURCES

■ **Case Study: Causes of Overpopulation in Bangladesh**

- Most land is delta land and is prone to flooding from the Ganges and Brahmaputra rivers.
- The changing courses of the delta distributaries prevent transport systems such as roads and bridges from being built.
- Birth rates are high and food production is unable to cope with population increases, owing to damage by flooding.
- Farm units are tiny and not capable of producing sufficient food for individual families.
- Transport is confined to unstable ferries that are routinely overloaded.
- Water supplies are untreated and contain pathogens (disease-carrying bacteria).

Consequences

- GDP is very low at only US$260 per person.
- Thousands die of contagious diseases, such as cholera and typhoid, because of water pollution.
- Thousands of people are left homeless after flooding.
- Thousands die near the coast owing to flooding on the flat delta coastal margins.

OR

THE DEVELOPMENT OF AGRICULTURAL LAND IN THE AMAZON BASIN

■ **Causes of Development**

- Poverty in the Sertao region of north-east Brazil is created by unreliable rainfall and drought.
- People were forced to abandon their homes in search of better land and a higher income.
- Families were willing to travel in groups to help each other make a new beginning.
- Government offers of free land in the Amazon region attracted settlers to the region.

- New trans-Amazonian highways allowed people easy access to the region that was not possible in the past.
- Felling of vast regions of forest by logging companies created huge ranches for MNCs to provide cheap meat for world markets.

Consequences

- Deforestation in the Amazon basin has taken place on a massive scale to create farmland.
- Tropical forest that absorbs carbon dioxide from the air has been lost.
- There is serious soil erosion by the tropical rains that wash away topsoil.
- Farm soil loses its minerals owing to leaching by the heavy rain; farms become useless and are finally abandoned for another patch of forest.
- Families do not increase their income and so remain poor.
- River pollution is created by mining activities for gold and other precious minerals.

3. The Influence of Society and Culture on Overpopulation

- Catholic and Muslim religions encourage high birth rates to increase their number of believers and spread their faith.

■ Case Study: Ireland

- There was overpopulation and subsequent migration from Ireland due to high birth rates. The population of the West soared between 1800 and 1845. Large families were unable to earn a living from the land. One million died and many emigrated.
- The remainder of the country was also overpopulated even up to 1962.

Consequences of overpopulation

- Low marriage rates occur as a consequence of migration, when young people leave the region.
- Small farms are abandoned or left in the care of older people, who are unable to work them efficiently.
- As population declines so does the region's political influence. The region increasingly becomes poorer.

■ Case Study: Brazil

- Parents had many children so some would survive to care for them in their old age.
- Many children died owing to a lack of sufficient food and medical services.
- Families were forced to migrate to other regions.

- The unequal division of land and the lack of education led to overpopulation in the Sertao. The unequal land division is a legacy from colonial times of the plantation system of farming.

■ **Case Study: India**

- Early marriage allows for a large family. Many young women marry as young as 14 years. This allows a long timescale for having children.
- There is a lack of choice about family matters. Many men want a male heir and lots of children to carry on the family genes.
- Lack of education and access to family planning prevents women from having control over their family size.
- Most women do not work outside the home. This leads to a lack of independence.
- The belief that the cow is a sacred animal leads to large numbers of malnourished cattle.
- Undernourished cattle in turn lead to low milk yields and poor farming methods, resulting in overpopulation.
- The tradition of having large families remains in a society even when incomes and education standards improve. This is leading to high populations in newly industrialised countries.

4. THE INFLUENCE OF INCOME LEVELS, EDUCATION, TECHNOLOGY AND POPULATION GROWTH ON OVERPOPULATION

- Large families of poor people encourage the belief that some will survive to cater for the needs of their parents when they are old.
- Poor farming communities lack the resources to improve their farming techniques, so yields remain low.
- It is believed now that even when incomes increase, families remain large owing to traditional practices. This accounts for the continued high birth rates and rapidly rising populations of newly industrialised countries.

■ **Case Study: Low Income in the Mezzogiorno in Southern Italy**

- Because land was owned by absentee landlords, southern Italy remained poor and failed to provide a decent living standard for its people.
- Until the 1950s many southern Italians were illiterate and this lack of education prevented them from improving their living standards.
- Millions were forced to emigrate or migrate to northern Italy to earn a living.
- Steep slopes prevented the mechanisation of large regions of the Mezzogiorno. This reduced farm output and helped to keep the region poor.

■ **Case Study: High Income in Japan**

- Even though Japan has one of the highest physiologic populations in the world, it is able to provide a high living standard for its people.
- A high education standard led to the development of electronics, engineering and computer industries that provide good incomes.
- The healthy diet practised by the Japanese comes from a high education standard and a long cultural tradition of eating fish.
- About 80 per cent of Japan's population live in cities. This increases income levels, as many jobs are provided in the services industry.
- Even though only 18 per cent of Japan's land is habitable and even less of it is capable of being farmed, it supplies a large amount of farm produce such as rice and vegetables.
- Huge investments have been made in research, in high-yielding varieties of rice and in innovative irrigation practices.
- Its fishing fleet accounts for 14 per cent of the world's catch and its aquaculture is very advanced, growing large amounts of fish produce in coastal bays and inland waters.

5. THE IMPACT OF POPULATION GROWTH ON RATES OF DEVELOPMENT

- Slowing population growth tends to make a population better off financially.
- When population growth falls less money is needed for hospitals and schools than would have been needed if population continued to grow. This reduces taxes and increases income.
- It does not follow that when income rises there is a corresponding fall in family size in newly developing countries, because of traditional and cultural influences.

ELECTIVE TOPIC 25
MIGRATION AND CHANGING MIGRATION PATTERNS IN IRELAND

Push factors force people to leave a region. They may include financial, religious, social or environmental reasons.

Pull factors attract people to a region. Again they may be financial, religious, social or environmental.

MIGRATION PATTERNS IN IRELAND

■ From West to East

- The Leinster region has increased its population each year since 1926.
- The percentage shares of the populations of Munster, Connacht and Ulster have fallen since 1926.
- People have migrated from Munster, Connacht and Ulster to Leinster since 1926.
- Connacht's population has fallen from 1.4 million people in 1841 to 433,000 today, a 70 per cent drop.

■ From Rural to Urban Regions

68 per cent lived in rural regions in 1926, so 32 per cent lived in cities.
54 per cent lived in rural regions in 1961, so 46 per cent lived in cities.
40.4 per cent lived in rural regions in 2003, so 59.6 per cent lived in cities.

REASONS WHY PEOPLE LEAVE THE WEST OF IRELAND FOR DUBLIN

1. Farms in the west of Ireland are small and unprofitable. People are leaving the land for jobs in the cities.
2. As young people leave the west, the services (e.g. schools, recreational centres and hospitals) close. The region becomes unattractive to live in, so even more people leave the area.
3. Industry is reluctant to set up in an area where the workforce is limited. So jobs are few and people leave to find employment elsewhere.
4 Standards of living are lower in the west than the east of Ireland. Young people leave the west for better lifestyles in the east.
5. Many industrial estates and business parks offer high-income jobs in the Leinster region.
6. Many young people attend third-level colleges and remain in the Dublin region, as they become accustomed to the higher lifestyle.

■ Effects of Migration on the West of Ireland

1. Many people between the ages of 18 and 30 migrate from the area, so marriage rates and birth rates are low.
2. Farms are left in the care of older people who often lack the energy to work them fully. Many farms become neglected or abandoned when the older people die.
3. Industry is reluctant to set up in an area of low population and out-migration, as the labour force is limited.
4. Community services and facilities decline as the population falls. This in turn reduces the attraction of the area for the young people.

■ **Effects of Migration on Dublin**

1. The population of Dublin has increased. Large suburbs and dormitory towns have developed around the city to create a greater Dublin with a population of 1.3 million.
2. Greater Dublin has expanded rapidly. Its commuter hinterland now reaches the Midlands to the west and Dundalk to Arklow in the east.
3. The cost of housing has risen massively, so that many people are unable to purchase their own home.
4. Young people from the west of Ireland help to develop the city's economy, e.g. bringing new skills, purchasing houses.
5. Overcrowding may be the result in parts of the city. Great demand for accommodation raises prices and this may lead to overcrowding.

POST-1950 MIGRATION TRENDS

■ **1950s**

- A lack of jobs led to a huge increase in emigration. 1958 had the highest net migration rate of 58,000 people.
- There was a net migration of 409,000 people in the 1950s.

■ **1960s onwards**

- Ireland's population reached its lowest level, of 2.8 million, in 1961.
- After 1961 Ireland's population began to grow again.
- There was reduced emigration during the 1960s, to a net low of 5,000 in 1971.
- Net in-migration from 1972 to 1979.
- High out-migration from 1980 to 1991.
- High in-migration from 1995 to 2002.
- Low in-migration from 2002 to 2004.

■ **Positive Effects of In-migration**

Cultural effects

- Foreign nationals make Ireland more outward-looking and cosmopolitan.
- People from China, Nigeria, the Balkans, India and the Philippines add a cultural diversity to an isolated, peripheral island nation.

Employment

- Many job vacancies are filled by people who are willing to work for lower wages, adding to the competitive nature of the job market.
- Foreign nationals bring new skills and ideas to the Irish workplace.

■ **Negative Effects of In-migration**

Repatriation of guest workers

- When foreign nationals are employed nowadays, it is generally on a fixed-contract basis.
- New migrants become used to a higher living standard than they were accustomed to at home.
- When their contract expires they must return to the prospect of unemployment and lower wages at home.
- Political pressure is often put on governments to allow such people to stay, especially if some of their children were born in the host country.

Refugees

- A high proportion of immigrants into some countries are refugees from wars or persecution. They occur generally in large numbers and the cost of coping with their needs can be a severe financial burden on the host country.

NEGATIVE EFFECTS OF EMIGRATION

- A skilled and educated workforce is lost.
- Many people from the ages of 18 to 40 emigrate. These ages make up the most energetic and youthful groups in a workforce.
- Emigration causes a brain drain that lowers the ability of a country to create indigenous industry.
- Loss of population reduces the home market, leading to a fall in sales, and this is reflected in a fall in production of consumer products.
- Peripheral regions are most affected by falling population, as they are the most vulnerable places where unemployment is most likely to occur in a period of depression.

MIGRATION POLICIES IN THE EUROPEAN UNION AND IRELAND

■ **Migration Policy in the EU**

- The EU allows for the free movement of workers among all EU countries. Its policy states, 'The mobility of workers must be one of the ways by which the worker is guaranteed the possibility of improving his living and working conditions and social advancement.'
- Migrant workers are entitled to remain in a country after working there. In principle this applies to refugees and EU citizens.
- The country that first accepts a refugee must take responsibility for awarding refugee status to that person.

■ Ireland's Immigration Policy

- The main components of Ireland's immigration policy include:
- Nationals from the European Economic Area (EEA) do not need a visa to live and work in Ireland. The EEA consists of the EU states plus Norway, Iceland and Liechtenstein. For all others, a visa is essential.
- Those who need a visa must apply for a work permit before they enter the state.
- Persons who claim asylum are given full-board accommodation while their claim is being processed, initially in Dublin and later at recognised centres throughout Ireland.
- Those who do not require a visa include:
  1. Persons who have permission to remain in Ireland, such as people with special skills and foreign full-time students.
  2. Persons who have refugee status.
  3. Persons who have been granted permission to remain on humanitarian grounds.
  4. Persons who are claiming refugee status while their claim is being processed.

ELECTIVE TOPIC 26
ETHNIC, RACIAL AND RELIGIOUS ISSUES CREATED BY MIGRATION

1. Race refers to biological inheritance. It refers to DNA or the genes passed from parents to children. There is no such thing as a 'pure' race.
2. Ethnicity refers to minority groups with a particular self-identity, such as Cubans in America or Chinese in Ireland.

Minority groups may be defined by:
- place of birth
- combined language and fertility rates, for example Hispanics
- religion, such as Muslim.

RACIAL DIVISION

■ Apartheid in South Africa

- Apartheid was racial separation of blacks from whites as a principle of society enforced by law.

- Before, under British law and after independence, segregation of blacks from whites was practised.
- In 1948 racial discrimination was justified and enforced by law.
- Nelson Mandela was imprisoned for protesting against this unjust law.
- Blacks were forced to live in poverty in 'homelands', the most deprived, almost uninhabitable regions of South Africa.
- All homelands people lost their right to citizenship of South Africa.
- Non-whites could not buy land.
- Apartheid ended in 1994.

■ **Ethnic Cleansing**

- This term was first used in the war that occurred during the break-up of Yugoslavia.
- It's a policy where ethnic groups are either slaughtered, or expelled by force, threat or terror from the country in which they live.
- The deliberate attempt to eliminate the Muslim people from Bosnia–Herzegovina led to ethnic cleansing.
- Examples include the massacres in Srebrenica and other enclaves.

■ **Religious Conflict in India**

- After independence in 1947, differences and fears arose between the Hindu majority and Muslim minority.
- Muslims believed that Hindus would have too much power over them once independence was achieved.
- Muslims demanded the partition of India along religious lines; the Hindus rejected the idea and riots broke out between the religious groups.
- India was finally divided into two states according to the religion of its people.
- Pakistan became Muslim and India became Hindu.
- Pakistan was in two parts, called West Pakistan and East Pakistan.
- Fighting continued after partition, so many Hindus in Pakistan fled to India and many Muslims in India fled to Pakistan.
- One state remained neutral for some time. Then it joined India and fighting between Pakistan and India over this region, Kashmir, continues to this day.

ELECTIVE TOPIC 27
RURAL-TO-URBAN MIGRATION

CONTRASTING IMPACTS OF RURAL-TO-URBAN MIGRATION IN DEVELOPED AND DEVELOPING REGIONS

■ **Developing Countries**

Why did cities grow in developing countries?
- The growth of cities in developing countries has been a result of population growth and of rural-to-urban migration before industrialisation occurred.
- So the cities came first, and then later industry developed over a period of only 60 years.
- Cities grew mainly because of rural 'push' forces, such as poverty and hunger.
- There was the hope of employment and the prospect of access to schools, health services, a safe water supply and other services.

Effects of rural-to-urban migration in developing countries
- There is an exceptionally high rate of population growth in cities, as most migrants are young adults of childbearing age.
- This young population accounts for over 60 per cent of urban population growth.
- Many cities have grown so large they are now called megacities.
- Squatter settlements, called shanties, or *bustees* in India and *favelas* in Brazil, have grown on the outskirts of cities.
- Too many people are being squeezed into cities that do not have the jobs to support them.
- Children are forced to earn a living as beggars, prostitutes and labourers.
- Males are the most likely to migrate to cities. This leaves an unbalanced male-to-female ratio in rural regions.

■ **Case Study: Bustees in Calcutta and Bombay**
- Over 60 per cent of Calcutta's population lives in bustees and over 500,000 homeless people live and sleep on the streets.
- Permanent slum dwellers are well protected by law in Calcutta.
- Pavement dwellers have no such rights.
- Many live under bridges, along canals or on land destined for other uses, such as roads.

- One in three people in Calcutta lives in a bustee.
- Open sewers are common and disease is widespread.
- Warm monsoon weather creates ideal conditions for malarial disease.
- Gender bias restricts young girls' educational chances.
- Employment chances are poor without education.
- Child labour is common.
- Population densities are four times higher than that of New York, i.e. 476/hectare.
- 77 per cent of all families have only one room in which to live.

■ Developed Countries

Why did cities grow in developed countries?

- Industrialisation occurred in small urban regions. The industries provided vast numbers of jobs that were then filled by migrants from rural regions.
- The industries came first, and then the cities grew gradually over 250 years.

Effects of rural-to-urban migration in developed countries

- 'Urban sprawl' is the term used to describe the rapid spread of urban regions out into the countryside. It consists of large housing-estate developments on the edges of cities.
- New growth centres develop some distance from the main cities. These growth centres are planned to cater for all the needs of their new inhabitants, including work and leisure.
- Migration to urban centres has caused depopulation of rural regions. As populations move away, the services of those rural areas close owing to lack of support.
- Farms are sold or left in the care of older people.
- Industry is reluctant to set up in rural regions because of the lack of a sufficiently skilled, educated labour force nearby.
- Migration to inner-city areas by illegal and legal new immigrants creates ghetto communities that have distinctive characteristics of their individual culture.
- Young, educated, unmarried, single people or divorced single migrants tend to live in urban-renewal areas of cities, for easier access to work, entertainment and friends.
- The daily movement of rural dwellers to and from work in the mornings and evenings creates rush-hour traffic.
- Counter-urbanisation occurs when the quality of life in a city deteriorates and some urban dwellers migrate to rural areas in search of a better quality of life.

■ **Environmental Problems Caused by Urban Expansion**

1. Global warming and the greenhouse effect

- Industry, people's homes and their cars are producing vast quantities of greenhouse gases that trap the heat that rises through the atmosphere.
- This build-up is causing the atmosphere to overheat.
- That in turn is causing melting of the Arctic and Antarctic ice sheets and a rise in sea levels.
- The United States, containing 5 per cent of the world's population, produces 25 per cent of greenhouse gases. About 82 per cent of these gases are from fossil fuels used to generate electricity and run cars.

2. What is smog?

- Smog is a combination of smoke and fog that hangs over a densely built-up area under calm atmospheric conditions. Because there is no wind to blow it away, the fumes from chimneys and car exhausts build up in the air.
- Smog may conceal a range of dangerous chemicals, such as sulphur dioxide and nitrous oxides, as well as a variety of gases from industrial plants.
- Old people, young children and those who already have lung disease are the most vulnerable to respiratory complaints from smog.

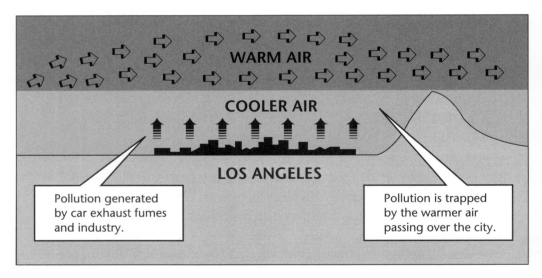

▲ Smog formation over Los Angeles

3. Urban waste

- Almost 30 per cent of urban sewage waste in Ireland goes untreated into inland rivers and lakes and into coastal waters.

- Sewage receives no treatment, or only preliminary treatment, prior to discharge in 53 (36 per cent) of the 390 urban areas studied by the Department of the Environment in 2000.
- Only 5 per cent has nutrient-reduction treatment.
- The Environment Protection Agency's Millennium Report found that sampling programmes for soil testing are mostly non-existent in local authorities, and few have implemented new management systems.
- Over €1 billion is being invested in waste-water treatment between 2000 to 2006 in Ireland.
- Sampling programmes for soil testing are mostly non-existent in local authorities.

ELECTIVE TOPIC 28

LOCATIONAL CHARACTERISTICS OF IRISH SETTLEMENTS

■ **What does the Site of a Settlement Mean?**
- 'Site' refers to the characteristics of the actual ground or point on which the settlement is located.
- 'Situation' refers to the location of the settlement in relation to its surroundings, such as other settlements, rivers and uplands.
- 'Location' includes the site and situation of a settlement.

THE SITE, SITUATION AND FUNCTIONS OF IRELAND'S SETTLEMENTS

■ **Prehistoric Settlements**

Ireland's first settlements
- The earliest Irish settlers were hunter-gatherers and belonged to the Middle Stone Age or Mesolithic Period.
- They came to Ireland about **9,000 years ago (7,000 BC).**
- They lived close to rivers or lakes where fresh water supplies were available.
- Many settled temporarily on coastal sites where heaps of shells and animal and fish bones were dumped in heaps called **middens**.
- Middens appear on Ordnance Survey maps in a linear pattern near present high-tide levels along the coast.

Ireland's first farming settlements
- These first farmers belonged to the Neolithic or Young Stone Age and the Bronze Age.

- They buried their dead in stone tombs called megaliths.
- Their tombs form a dispersed pattern:
  - across the Burren in Co. Clare
  - in the drumlin lands of Sligo to Dundalk
  - in West Cork, where the people mined for copper.
- They chose upland areas and raised, dry or hilly lowland sites because the gritty soil was easier to till than the heavy clays of lowland areas.
- The Young Stone Age settlers came about **7,000 years ago (5,000 BC)**.
- The Bronze Age settlers came about **4,000 years ago (2,000 BC–650 BC)**.
- Their works include **megaliths, stone circles, cairns, cist graves, standing stones, wedges, fulachta fiadh.**

Celtic settlements

- These farmers belonged to the Iron Age, from 650 BC to 250 AD.
- They introduced iron working to Ireland.
- They built their homes in a dispersed pattern throughout farming lowlands.
- They divided the country into tuaths.
- They built **hill forts, ring forts, crannógs, cahers or cashels (stone forts) and promontory forts** that were built on cliff edges for protection against attack.
- Elevated sites were generally chosen for the more important larger settlements.
- Ring forts are often referred to as lis, or dun, and are sometimes recorded in place names such as **Lisdoonvarna, Liselton** or **Dunquin**.

Small monastery settlements

- Individual missionaries chose isolated sites in glaciated river valleys for their settlements.
- Remains of their stone churches are mostly all that remains of their settlements.
- These sites are listed on maps as **ch,** printed in red. Local place names that include the word 'kill' often suggest such settlements.
- These settlements were near streams or lakes for a freshwater supply.

THE HISTORIC DEVELOPMENT OF IRISH TOWNS

■ Large Early Christian Settlements

1. These were sited at
   - route centres such as Clonmacnoise in Co. Offaly
   - on fertile plains such as Kells in Co. Meath.
2. They were centres of religion and education for people from Ireland, Britain and the European continent. At the centre was the monastery with its churches, round tower, monks' dwellings and graveyard.

3. The words Manister, Monaster, Kil, Cill or Ceall on Ordnance Survey maps all suggest that the town developed as a monastic centre.
4. As the towns grew they developed other functions, such as markets.
5. All routes focused on the centrally placed monastery, creating a radial route pattern.
6. The Normans often took control of these settlements. They constructed defensive walls and new buildings nearby, and added more functions.

■ **Viking Settlements**

- These settlements were built at defensive sites on sheltered river estuaries.
- They also acted as trading bases or ports for trade.
- They include Dundalk, Drogheda, Dublin, Wexford, Waterford, Cork, Tralee and Limerick.

■ **Norman Settlements**

1. The Normans came to Ireland in 1169 and spread west and north, capturing the best farmland and building castles and towns to protect this captured land.
2. They choose good defensive sites, initially for mottes, and later for castles either:
    - beside existing thriving monastery settlements or
    - on new sites that were easily defended.
3. They chose:
    - bridging points inland and lowest bridging points on coastal estuaries, and
    - river loops, islands and elevated sites.
4. Unplanned towns developed around the castles, which were enclosed within high, defensive walls with guarded gateways for protection.
5. Abbeys, priories, and friaries were generally built outside the town's walls.
6. Abbey farms, called granges, indicate Norman origins for settlements.
7. Abbeys provided services such as education, accommodation for travellers, alms for the poor and hospitals for the sick.
8. Norman towns were market centres, where fairs and markets were held at regular intervals.
9. Milling and tanning industries developed in these towns.
10. Cas, castle, motte, town wall, gate, town gate, abbey, friary, priory, grange, castle land: all these indicate Norman origins and are printed in red on Ordnance Survey maps.

■ **Planned Plantation Towns**

- Planned towns were built as part of the plantation of either Laois–Offaly, Munster or Ulster.
- All have parallel or evenly wide streets.
- Centrally placed Protestant churches introduced the new faith at that time.
- The towns had central diamonds or squares where markets and fairs could be held.
- Later planned towns of the 18th and 19th centuries had wider streets than early plantation towns.
- They had tall, three-, four- or five-storey buildings along their streets.
- The Market House or building was located in the square for commercial purposes.
- A castle in or near the town was where the local landlord lived.
- Large demesnes or estates attached to the town were owned by the landlord and sometimes surrounded his manor house or castle.
- Demesnes or estates were surrounded by low stone walls and a ring of deciduous trees.

■ **Coastal Cities and Towns**

- These towns form a linear pattern along the coast.
- They were Viking, Norman or plantation towns that served as ports and/or market centres for large rural hinterlands.
- They were the focal points of inland and coastal routes.
- They had brewing, milling, warehousing, religious, educational and administrative functions that developed over time.

■ **Canal Towns**

- These towns include Newry, Mullingar, Tullamore and Athy.
- Rivers in the Midlands and east were widened and deepened.
- Canals were built during the eighteenth and nineteenth centuries to carry bulky goods from our largest cities and ports to inland towns.
- Many of these towns were already established on inland riverside sites before canal construction.
- Grain and timber were brought from inland towns to the ports for export. Beer, coal and other imported goods were brought from ports to the inland towns.
- Canal barges carried people.
- Large grain stores and mills, warehouses and hotels were built alongside canals for easy access for barges.

- Mills for grain and flour and for wool and linen were built in many towns on canal routes.

■ Railway Towns

- Railways were built in the nineteenth century.
- Hotels were built near railway stations to cater for long-distance travellers.
- Towns expanded as a consequence of this new business.
- Cattle were transported all over from fairs to other towns and especially to the ports.
- Railway towns prospered at the expense of canal towns. Trains were quicker and cheaper than canal transport.
- Commuting has once again given new life to these railway towns, as many people travel to Dublin by train to reduce car expense and avoid long traffic delays.
- Trains are also safer than road transport.

■ Seaside Towns

- Seaside towns developed as railways were built to join them to nearby cities.
- A rise in disposable income and a general improvement in people's standard of living allowed them to spend some time near the sea.
- Army camps were built for summer training of part-time military volunteers at seaside locations. This encouraged the development of golf courses nearby.
- Hotels developed near seaside beaches to cater for visitors.
- The layouts of these towns were planned with wide, straight streets.
- Wealthy people built holiday homes along the seafronts to take advantage of the views.

■ The 18th- and 19th-Century Expansion of Large Irish Cities and Towns

1. Most urban rebuilding and redevelopment took place during the 18th and 19th centuries.
2. This period of urban growth was called the Georgian Period.
3. Wide streets formed a mesh with blocks of buildings in between, creating a grid pattern.
4. Many tall five-storey buildings were built of red brick, with large windows.
5. They formed the Georgian suburbs at the edge of the old, unplanned, medieval towns.
6. Today parts of these suburbs form our towns' commercial districts.

■ New Towns

- Towns such as Shannon New Town in Co. Clare, and Tallaght and Blanchardstown in Dublin, were built to cater for the many rural migrants who came from rural and inner-city regions.
- Shannon was a well-planned town with local services and industrial estate to cater for workers' needs.
- Tallaght and Blanchardstown grew from village size to major urban centres within a few decades, but until recently lacked the services that normal urban areas provide.
- They form part of greater Dublin and are satellite towns of Dublin.

ELECTIVE TOPIC 29
RURAL SETTLEMENT PATTERNS

There are three categories of rural settlement pattern.

DISPERSED OR SCATTERED PATTERN

- This is created by widely spaced homes.
- Dispersed or scattered housing is usually associated with farmhouses with outbuildings or sheds nearby. The pattern developed when farms were enclosed after commonage-type farming was abandoned.
- Farm buildings are widely scattered where farms are large, such as in the rich farmlands of Counties Meath, Westmeath, Tipperary, Limerick and Clare.
- In more western regions, where farms are small, they are more closely spaced but still scattered. Many farmhouses are located at the end of long passageways or on roadside sites.
- Since the 1960s many non-farmers who are also rural dwellers built one-off houses where sites were available for purchase. These homes are either single- or two-storey dwellings generally on roadside sites.

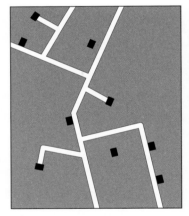

▲ Scattered or dispersed housing is common in rural regions

CLUSTERED SETTLEMENT

- This pattern is created by groups of houses.
- Dwellings that are grouped together are generally farm dwellings of the 18th, 19th and 20th centuries; or in isolated cases they may be remnants of the Clochan system of the west of Ireland.
- Farmhouses were built in clusters in Counties Kilkenny and Waterford as part of the division of land in the 18th and 19th centuries.
- Some clusters were built at road junctions where shops and post office, and maybe a filling station and shop, have developed over time.

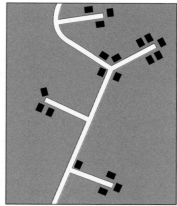

▲ Buildings are generally in a cluster in some rural regions

RIBBON SETTLEMENT

- This is a recent pattern development. It is generally composed of individual, one-off houses that developed along roadways. Available building sites and a desire by rural dwellers for new, fashionable houses with privacy on their own large, individual site created this pattern.
- Local planning authorities were lenient as regards planning permission, and there were no overall planning controls for such housing from the 1960s until 2000.
- The presence of telephone cables, electricity lines and piped local authority or private water schemes also encouraged this kind of development.

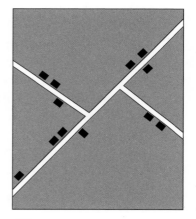

▲ Buildings are located on roadside sites

- Landowners could also increase their income from sales of individual roadside sites and obtain high prices for them.
- Suburban arterial routes (main roads) were the preferred choice for filling stations, bed-and-breakfast accommodation and buildings for local people. In recent years the increase in road traffic has restricted the development of dwellings on such routes.
- County and national development plans do not encourage ribbon development any more, as it is regarded as unsustainable.

ELECTIVE TOPIC 30

THE EXPANSION OF CITIES AND PRESSURE ON RURAL LAND USE

- As cities expand they force surrounding rural areas and some inner-city areas to change their land-use function from green areas to built-up, urban land uses.
- A megalopolis is a cluster of large cities that have expanded and joined, creating a vast, urban environment. Imagine an urban region the size of the entire county of Cork.
- The Randstad is a megalopolis in the Netherlands that has endangered its Greenheart, an open green region of farmland with scattered small towns and villages.
- **Green belts** are open spaces of rural land use, parks or woodlands within towns or between towns.
- Green belts are designed to be permanent features to:
  1. prevent the development of vast urban regions
  2. prevent continuous urban environments
  3. provide recreation areas within urban centres.
- Local communities, as well as local government, should develop strategies together to protect and manage these areas successfully.

ELECTIVE TOPIC 31

RESIDENTIAL AREAS AND SEGREGATION WITHIN CITIES

People with different incomes, cultural or racial backgrounds tend to cluster in separate parts of a city.

LOCAL GOVERNMENT HOUSING

- Certain environments attract a particular type of housing. Local authorities provide subsidised housing, either in flat complexes in inner-city areas or in housing estates.
- Many housing estates, such as those that were built in the 1960s and 1970s, have led to no-go areas where lawlessness is rampant.
- Residential groups sometimes interact with developers and planners to produce areas with compatible neighbours with whom they have most in common.

GHETTOS

- A ghetto is an area of a city that is settled by a minority racial, religious or national group with certain characteristics that distinguish them from the urban population as a whole.
- The term ghettos originally referred to sections of European cities where Jews settled or were forced to live.
- Ghetto today refers to areas where black and other minority groups live, e.g. Chinatown or Harlem in New York.
- The term ghetto is a product of discrimination by society against a certain group.
- People of a particular underprivileged group live together for support.
- Cities that experience high immigration tend to be structured in a series of concentric zones of neighbourhoods of different ethnic groups.

ELECTIVE TOPIC 32
URBAN PLANNING AND URBAN RENEWAL

IMPROVEMENTS CREATED BY PLANNING AND RENEWAL

- Old buildings and derelict sites have been renewed, creating a vibrant, young city that attracts shoppers and nightlife.
- New streets create easier traffic flows.
- Parking zones and multi-storey car parks
- The restoration of old buildings with architectural character
- Pedestrianised streets and new pedestrian crossing places
- Disc parking that contributes to improved traffic flow and revenue
- Ring roads and bypasses
- Tunnels under river estuaries and channels, e.g. the Jack Lynch tunnel in Cork and the Dublin Port Tunnel.

■ **The Dublin Transport Strategy**

This plan involved:
- a vision statement to create a plan and vision for the future
- an integrated public-transport system that can be reached within a 10-minute walk at most
- quality bus corridors
- a light rail system in Dublin
- cycle routes

- a National Roads Authority that has responsibility for the development and management of our roads. This includes:
  1. PPPs (Public–Private Partnerships)
  2. toll charges.
- PPPs may involve toll charges on new developments over a 30-year period to recoup costs and maintenance charges for investors and obtain best value for money for the taxpayer.

ELECTIVE TOPIC 33
PLANNING STRATEGIES IN RURAL AREAS

SUSTAINABLE DEVELOPMENT

This involves environmentally friendly planning that seeks an acceptable quality of life for present and future generations.

It is the careful management of economic activities so that local environments and people's activities are interdependent.

Planners try to promote orderly development to:
- ensure the land is used for the common good (the good of everybody)
- meet the needs of society for housing, food and materials, employment and leisure
- support policies concerned with regional development, social integration, urban renewal and the maintenance of strong rural communities
- balance competing needs and protect the environment as much as possible.

Careful planning can help to achieve these objectives in a number of ways:
- by controlling the development of transport, natural resources and the efficient use of energy
- the careful location of industry, houses and business/shops/services
- by controlling the shape, size and structure of settlements
- effectively using already-developed areas.
- protecting and supporting our natural environment and wildlife habitats, including areas and features of outstanding beauty.
- accommodating new developments in an environmentally sustainable and sensitive manner
- strengthening villages and towns, both socially and economically, in order to improve their potential as growth centres.

■ **County Councils are Legally Bound**

They are obliged to:

1. determine a policy for proper planning and development
2. implement the National Development Plan by:
   - controlling planning and enforcing planning decisions
   - creating sustainable development in rural areas that respects nature, natural systems, natural habitats and species and protects the environment
   - making good-quality decisions and encouraging public participation, openness and proper enforcement
   - being responsive to change and reviewing development through compulsory five-year reviews.

ENVIRONMENTAL ISSUES

Urban-generated housing in rural areas is regarded by the National Development Plan as being unsustainable because these houses:

- are isolated and away from central services
- are serviced by septic tanks that may pollute ground water
- create suburban development.

Some disagree because they believe that new rural housing has some advantages, such as:

- a healthier rural environment than cities for family life
- the need to sustain rural community life and support services.

■ **Environmental Impact Assessment (EIA)**

This involves compulsory environmental impact assessments for:

- major developments such as new roads, large forestry projects that exceed 70 hectares
- the location of waste-material disposal sites
- projects that do not reach acceptable levels of agreement or standards, if it is believed that the project would affect the local environment negatively.

■ **Strategic Environmental Assessment (SEA)**

SEAs examine the policies, plans and programmes of environmental impact assessment.

ELECTIVE TOPIC 34
URBAN HIERARCHY, HINTERLAND AND CENTRAL PLACE THEORY

- Settlements can be classified according to size, function and population density.
- A major function of all settlements is to provide services for their inhabitants and the people who live in their hinterlands (surrounding areas).

Central Place Theory tries to explain that the arrangement of towns is determined by the hinterlands that they serve.

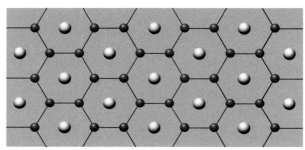

◀ Model of hexagonal areas

Hexagonal areas
This produces no competition and leaves no area unserved so it is the best model.

- ● first-order (lowest) settlement, e.g. village
- ◑ second-order settlement, e.g. town

THREE BASIC CONCEPTS OF CENTRAL PLACE THEORY

1. The range of goods is divided into three categories: high-order, medium-order and low-order goods and services.
2. Frequency of demand refers to the level of demand for goods and services, such as daily needs, weekly needs, monthly needs or annual needs.
3. Threshold refers to a certain threshold, or minimum number of people, required by each shop/service to be viable. Some shops/services have higher thresholds than others to remain profitable or at least cover their costs: for example, the threshold for a supermarket will be different from that for a local shop.

   ■ Hinterland or Trade Area
   - Every city, town or village has its own hinterland or area from where people travel to do their shopping. Cities have large hinterlands and villages have small ones.

- Hinterlands may be affected by physical features such as unbridged rivers, mountains or high upland, bogland or areas liable to flooding, which may distort and reduce their almost ideal circular shape.
- Hinterland size also varies according to density of population. Villages in isolated regions may have large hinterlands, such as in the west of Ireland, to remain as viable communities; whereas in high-density lowlands a number of villages may each be viable in a similar-sized region.

■ How Modern Changes Affect the Number of Functions

- Modern transport, such as cars, allows people to travel further and bulk buy. This affects the range of services that are viable in rural villages.
- Deep freezes and convenience foods reduce the need to make daily trips to local shops.
- Larger settlements can be more competitive and by even lower prices can attract more customers.
- Villages may no longer have sufficient populations to support their traditional functions, leading to the closure of many of these service outlets.

■ Some Criticisms of Central Place Theory

- It was designed to work on a featureless plain that does not really exist in nature.
- Modern transport systems have undermined the original concept, as they favour some centres more than others.
- Population is not evenly dispersed.
- Settlements compete with each other to enlarge their hinterlands.

ELECTIVE TOPIC 35
CHANGING URBAN LAND-USE PATTERNS

LAND-USE ZONES AND LAND VALUES IN CITIES OF THE DEVELOPED WORLD

- A city's land uses may be divided into concentric zones. The oldest is at the centre and the youngest is at the city's edge.
- The oldest parts at the centre are often the present-day commercial downtown districts.
- The city centre is surrounded by a band of old housing with some old light-industrial sites that may now be derelict sites or renewed structures.

- These old housing regions may house ghetto communities.
- A band of newer housing or pockets of high-income housing may surround the old housing.
- The newest housing is in the suburbs, on housing estates.
- Heavy manufacturing is now located in industrial estates on major routes.
- Many office services and wholesale outlets are located in business parks.
- Shopping complexes and hospitals create growth centres in certain locations on city boundaries.
- Some wholesale and light manufacturing land uses form wedges or sectors along major routes, increasing with distance from the centre.

■ Central Business District

- This is the heart of the city, with department stores and specialist shops.
- It has the highest land values and tallest buildings.
- There are multistorey buildings, offices and apartment blocks.
- Financial and commercial land uses are the most common.

■ Industrial Zones

- The first industries in urban centres were located close to the city centres, because the towns were small. Almost all of these are now closed and their sites are taken by new apartments, shopping or office complexes.
- New industries are located in industrial estates on the outskirts of towns and cities and on main routes.
- Heavy industries are located close to water routes for easy import and export of goods.
- Business, wholesale and science parks are also located on the edge of cities or towns.

■ New Suburban Downtowns

- A new type of growth area involves suburban downtowns in large urban regions.
- Offices, hotels, department stores, industrial parks, entertainment facilities and car parking are grouped together to create a growth centre.

LAND-USE ZONES IN DEVELOPING WORLD CITIES

■ The Central Business District

This will have:
- the business, employment and entertainment centres
- a central square or plaza with government buildings

- a spine of commercial land use surrounded by high-class residential housing radiating out from the core
- sectors of the best housing around the core
- sectors of modest housing and derelict sites surrounding the high-class housing
- shanty towns surrounding everything for many miles.

ELECTIVE TOPIC 36
THE FUTURE OF URBANISM

Cities today are places:
- composed of different parts, with culturally mixed and multi-centred urban areas
- where intense connections can exist with far-off places but frightening disconnections between neighbourhoods
- where past assumptions that far-off people and places may not relate, while nearby ones do, are not necessarily true
- where stark contradictions and huge tensions coexist
- where highly planned and expensive developments that took planners and skilled consultants many years to put together exist in some areas, while other parts of the same city appear to be uncared for.

THE DUAL CITY
- This name refers to social division within urban areas.
- A large proportion of a city's population is cut off from job opportunities that would help them improve their quality of life.
- Most of these jobless people are from ethnic minorities, are elderly or disabled people, or are single parents.
- Many 'jobless' people make a living from sources, sometimes illegal, that pay in cash, and are not known to the revenue commissioners.
- It is acceptable today to create regions in cities that form distinct social groups, creating alienation of some communities that are not socially acceptable.
- Isolating urban communities creates a fear among the rich and despair among the poor.

■ **Urban Renewal and Mega-projects**

Urban renewal

- removes low-income housing for very-low-income homeless people
- creates forced out-migration of low-income rental groups or individuals.

Mega-projects

- These introduce specialised and exclusive living and entertainment environments for high-income groups.
- Many mega-projects are funded through tax incentives and zoning.
- Many cities try to promote themselves by creating a certain image that is attractive to the young.
- Urban industry promotes itself though images of business parks, landscaped lawns and clean environment.

■ **Public Space and Social Control**

- Freedom of movement within cities is becoming more restricted. Only those who belong are welcome in some areas.
- Surveillance cameras monitor people's movements and activities.
- Cameras are sometimes used to exclude poor groups from some areas, such as high-priced shopping areas.

The desired advantages of CCTV

- significant reduction in crime within CBD areas
- creates a feel-good factor for shoppers
- revitalises town centres through a sense of security for consumers, thereby creating a spending environment.

The development of neighbourhoods

- Successful urban development involves encouraging development at neighbourhood level.
- It also means rejecting blueprints from outside.

SECTION 3:

OPTIONS
(HIGHER LEVEL ONLY)

Higher level students must study one of the following:

Option 1: Global Interdependence pages 204–242

OR

Option 2: Culture and Identity pages 243–272

OR

Option 3: Geoecology pages 273–292

OR

Option 4: The Atmosphere–Ocean Environment pages 293–311

NB: Study only <u>ONE</u> of these options.

OPTION 1: GLOBAL INTERDEPENDENCE

OPTION TOPIC 1
MODELS OF DEVELOPMENT

THE MEANING OF DEVELOPMENT

- In the past development referred to
  - the state of a country's economy, which was judged solely by the GNP of an individual, or
  - the average wealth produced for an individual by the country in one year.
- Today it is felt that real development must include sustainable economic, medical, spiritual and cultural aspects of a society.

Development, according to Abraham Maslow, includes a number of human needs that must eventually be reached before a society can be defined as developed. They include:
  - basic needs, such as clean water, balanced diet, access to good healthcare
  - security, such as personal protection from violence by any individual, groups or the state
  - being valued by society, loved by family and friends
  - self-respect, through secure employment, with adequate income
  - personal growth through development of a person's talents.
- The last few needs can be achieved only when basic needs are fulfilled. So there is a 'ladder' of human needs.

■ Differing Development Models

1. National self-reliance

- The aim of a country is to produce sufficient products for sale abroad so that export earnings could be balanced against the cost of imports. To achieve this the state would take over or nationalise important industries and practise protectionism to support home industry.

- The profits from state-owned industries were used to improve living standards, education and health.

But protectionism led to poor work practices and inefficiency:

- National markets were too small for continued development.
- It was difficult to penetrate foreign markets owing to the high cost of home-produced products.
- The cost of energy supplies such as oil (during what was called the 'oil crisis') soared, creating huge debts that absorbed all the profits from home industries.
- Developing countries sought help from the World Bank and the International Monetary Fund to ease their repayments.

2. **Centrally planned development**

This economic model was based on the ideas of Karl Marx. It was practised in the former USSR and China, and is still practised in Cuba and China.

- The aim of this policy was to create greater equality in the distribution of wealth and the provision of at least the basic needs of every person.
- To achieve this all factories, banks, and land were nationalised and managed by the state.
- The economy's profits were reinvested in the state to provide essential services and industrial products that could be sold or used in further production.
- There were great social gains in education, healthcare, sport and equality for women.
- In the socialist-controlled Kerala region of India, over ninety per cent of girls and boys receive a formal education and life expectancy is 19 years greater than in the rest of India.
- But the system discouraged individual initiative.
- No alternative existed to the single governing party.
- Strict government controls restricted private enterprise.
- Lack of competition at home led to inefficient and shoddy work practices.
- The system failed to produce sustained economic growth.

3. **Modernisation and free trade**

This model is designed to create a world economy (globalisation) where all barriers to trade are removed and private enterprise (capitalism) is unrestricted by governments or people.

- The manufacture or creation of exports for sale in a world market means profits can be used to purchase imports or pay debts.
- Free international competition will create sustainable economic growth.

- This has resulted in significant improvements for small economies such as Ireland.
- Third-world countries such as South Korea and Brazil have also gained significantly.
- It is a capitalist system based only on profit.

But this model has its disadvantages too:

- Profits are valued above everything and everybody.
- In some countries sweatshop production and abuse of low-paid female and child labour are used to increase profits.
- Vast wealth is created for a tiny few and so is very unevenly distributed.
- Some services such as hospitals, public transport and water supplies are provided by private enterprises so they be run more efficiently for profit. The introduction of this model in some countries, like Brazil, caused unemployment when native industries closed, as they could not compete on a global scale.
- Local currency was devalued to make exports easier but imports, such as oil, became more expensive.
- Governments could not subsidise basic foodstuffs and so created hunger and undernourishment.
- Cash crops were encouraged to pay foreign debt, at the expense of using land to produce foodstuffs for local consumption.
- All countries from all traditions are encouraged to follow this Western model of Americanisation.
- The wealth and economic development of some societies are based on the exploitation and poverty of others.

■ Examples of Exploitation and Underdevelopment

- In the 17th and 18th centuries India was one of the world's most economically developed economies.
- Under British control native industries were discouraged and India became a provider of raw materials for British industries.
- The Incas of the Andes had more advanced agricultural practices than Europeans at that time. Terracing, irrigation and other methods were abandoned as they became slaves for their conquerors while providing raw materials for Spain.
- Many sub-Saharan peoples lost their lands to Belgian, French, British and other colonial powers. Much of their land was used to grow luxury crops for the colonial powers, rather than food for their people.

Large multinational companies transfer vast sums in profits to their base countries. Much of this wealth is generated by the low-paid workers of third-world nations.

OPTION TOPIC 2
IMAGES AND EUROCENTRIC THINKING

- Europeans believed in the notion that they were superior to those people of the European colonies. The colonists referred to them as 'natives', as in 'savage natives'.
- Magazines and newspapers reflected this view in the images they presented to their readers.
- The cultures and languages of the colonies were looked upon as inferior to those of the Europeans.
- They believed that Europe was the centre of the world, so their maps placed Europe in the centre.
- Racist attitudes were developed towards colonised people.
- Colonial languages were encouraged in the colonies at the expense of the local languages.
- Europeans believed that the European idea of a civilised world was the ideal role model for all people everywhere.
- World maps placed Europe at the centre while countries such as China, with a much older and more advanced culture, were placed at the edges.
- The methods chosen to represent the different continents also exaggerated the size of Europe relative to the other continents.
- This belief that Europe was the best is called **Eurocentric Thinking**.

THIRD-WORLD IMAGES IN FIRST-WORLD COUNTRIES

- The underdeveloped countries contain two-thirds of the world's population, but they receive only one-tenth of news time on Irish news bulletins.
- Most news from third-world regions focuses on wars, famines and disasters.
- Local people are stereotyped as inactive, helpless victims, rather than creative and willing participants.
- TV often focuses on trivial news items when larger, more important issues need to be aired.
- Few news programmes examine the real causes of underdevelopment, such as unfair prices or unfair trading practices.

- Few local people are interviewed on programmes that concern their countries.
- Images from NGO agencies appeal to people for charity. These images may have a reduced effect over time. They also tend to reinforce a stereotyped image of third-world people.

OPTION TOPIC 3
EXAMINING WORLD MODELS AND LABELS

The poorest regions of the world are referred to in a number of ways.

Up to the 1980s the world was split into three divisions:

- The First World represented the rich regions such as Western Europe, North America, Australia and Japan.
- The Second World represented the communist countries of the USSR and Eastern Europe, such as Poland and Hungary. These people enjoyed adequate living standards.
- The Third World represented the 75 per cent majority of the world's people who were poor.

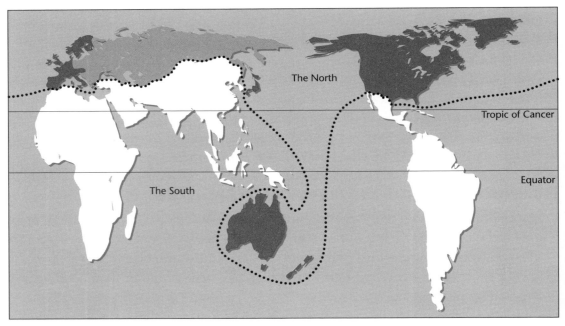

▲ This map is called Peter's Projection. It shows the correct sizes but not the correct shapes of continents. How does this view of the world differ from that presented by most maps of the world that are used in Ireland?

Objections to this categorisation

The Three-world Model

1. It was felt that this created a three-tiered society with the rich countries at the top.
2. The Second World no longer exists, owing to the fall of the communist system in the former USSR.
3. The term Third World suggested a third-rate world. However, geographers use the term to suggest the lack of real political power enjoyed by these nations.

■ The Two-world Model

The Brandt Report in 1980 suggested the following division:

*   **The North,** representing the rich and powerful countries. This includes the First and Second Worlds. However, not all of the countries are located in the northern hemisphere: for example, Australia is not.
*   **The South,** representing the poor countries of the Third World. However, there are great differences in stages of development among these countries.

■ Other Suggested Models

*   **Developed countries,** representing those where industry and services are well developed and people have a good living standard economically.
*   **Quickly developing** countries, representing those where industrial development is quickly being established and is leading to improved living standards.
*   **Slowly developing** countries, representing those that are still without any real industrial development and remain the poorest regions of the world.

OPTION TOPIC 4
SOME IMPACTS OF MULTINATIONAL CORPORATIONS (MNCs)

Multinational corporations are also called Transnational Corporations (TNCs).

■ Multinationals:

*   provide large sums for investment
*   provide large numbers of jobs
*   provide branch plants for manufacturing and research
*   increase exports

- increase imports for manufacturing
- create global trading networks.

■ **But Multinationals also:**

- cause job losses when local firms close as they are unable to compete with large companies
- cause job creation in one region at the expense of job losses in another
- cause branch plants to close because of decisions in another country
- work on the principle of profits first.
- The power of some multinationals can undermine the rights of workers to form trade unions, or can undermine the government by threatening to withdraw all investment. Many multinationals are wealthier than some countries.
- Multinationals return millions of their profits that they make in foreign countries back to their home country.

■ **Case Study: Wyeth Nutritionals in Askeaton**

- This is Ireland's largest producer of infant formula foods.
- It's a branch factory of the American Home Products Corporation.
- More than 600 people are employed directly at Askeaton.
- Other service jobs are provided indirectly in transportation and engineering.
- Wyeth contributes more than €150 million annually to the Irish economy.
- As Wyeth Askeaton expanded other plants closed, such as their plant at Havant, in southern England, and other factories in Australia, Colombia and South Africa.

■ **Effects of Multinational Exports from Ireland**

- Imports to underdeveloped countries must be paid for by profits from the exports of that country. If there is an imbalance then money must be borrowed to offset the cost.
- Imports may be sold at exceptionally low prices to eliminate competition from other factories operating in that country. This may negatively affect indigenous industries.

■ **Social and health effects**

- Breast-feeding is the most desirable form of infant feeding.
- Third-world babies are twenty-five times more likely to die if they are bottle fed.
- It is impossible to keep bottles and teats sterilised in regions that have low hygiene standards.

- Powdered milk for babies is expensive and can lead to family malnutrition.
- The World Health Assembly has banned direct advertising of milk-formula food in Third World countries.

OR

■ **Case Study: Kerry Group plc**

- This is an Irish multinational company.
- It started in 1972, with 40 workers and a turnover of €1.3 million in its first year.
- Today it employs over 20,000 people and has an annual turnover of €3.7 billion, with 75 branch plants in over 16 countries.
- It has markets in over 120 countries and trades with 35 of the top 40 global food manufacturers.
- Demand continues to grow for convenient quality foods.
- The Kerry Group are constantly sourcing their raw materials where prices are lowest.
- Kerry Ingredients is the largest division of Kerry Group plc and is one of the world leaders in the global food ingredients and flavours market.
- It employs over 400 food scientists, producing over 10,000 products.
- It has 30 processing plants in Ireland.
- It processes over 120 million gallons of milk in Ireland each year.
- It has manufacturing plants in every continent except Africa.
- Its largest sales are in Canada, the USA, Mexico, Brazil, the EU and Australia.

OPTION TOPIC 5
DEFORESTATION, GLOBAL WARMING AND DESERTIFICATION

We live in an interdependent, global economy. Actions taken in one area have an impact on other areas.

Three issues will be dealt with in this chapter. They are:
- deforestation
- global warming
- desertification.

Each of these processes is either a cause or an effect of processes elsewhere.

DEFORESTATION IN THE AMAZON BASIN

Most of the Amazon Basin in Brazil is covered by forests called 'selvas'.

The region contains one-third of all the tropical forests on Earth.

Up to the 1960s this region was lightly populated with native American Indian tribes, who lived as hunter-gatherers in the forests.

■ The Causes of Recent Exploitation of the Forest's Resources

- These selvas are viewed as a rich source of tropical wood.
- They are seen also as a wilderness region to be conquered to create profit.
- They have vast mineral deposits of iron ore, bauxite, gold, silver and tin, timber and oil.
- Most of the rich east-coast farmland is owned by landlords.
- The Amazon Basin was seen as a way to give land to the poor, hungry, landless peasants without causing conflict with the powerful landowners.

Cattle ranching

- Large beef cattle ranches aided by government funds focus on producing low-cost meat for American fast-food outlets.
- Many ranchers used the minerals in the soil to provide grass for the animals. But due to the heavy equatorial rains the soil quickly lost its fertility and the land turned to scrubland after 8 to 10 years.

International financial institutions

- The World Bank supported large projects like the Polonoroeste Project, which was responsible for large-scale deforestation of the selvas in the state of Rondonia.

■ The Effects of Deforestation

- Forest peoples are being forced from their natural environment by the cutting down of the forests.
- These tribes traditionally lived by hunting, fishing and subsistence.
- Constant contact by 'outsiders' is eroding their culture and bringing deadly diseases, such as measles, that their immune systems are unable to withstand.
- Traditional social life has been shattered. Many of the survivors are forced to live in squalid roadside conditions.
- One-quarter of all medicines owe their origins to rainforest plants, even though only one-tenth of these species have been studied.
- Selvas thrive in a very sensitive, balanced ecosystem. Deforestation upsets this balance and leads to a series of knock-on effects.
- Soils that are exposed to the heavy tropical rain are quickly washed away.

- **Global warming**: Trees naturally absorb carbon dioxide from the atmosphere and emit vast amounts of oxygen back into the atmosphere by day. When trees are cut down less oxygen is reproduced, the amount of carbon dioxide increases and the balance of the atmosphere changes.
- As carbon dioxide retains heat in the atmosphere it leads to global warming and rapid climate change.
- **Desertification**: As a consequence of global warming, some desert regions such as the Sahara in North Africa are expanding.

GLOBAL WARMING

- The Earth's atmosphere is gradually getting warmer. Glaciers are melting in mountainous regions such as the Alps and the Andes.
- 1999 was the warmest year of the twentieth century.

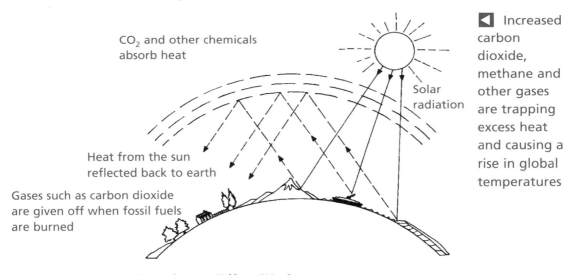

CO₂ and other chemicals absorb heat

Solar radiation

◀ Increased carbon dioxide, methane and other gases are trapping excess heat and causing a rise in global temperatures

Heat from the sun reflected back to earth

Gases such as carbon dioxide are given off when fossil fuels are burned

■ **How the Greenhouse Effect Works**

1. The Sun heats the Earth's surface.
2. The Earth's surface radiates this heat back into the atmosphere as long-wave radiation.
3. The normal amounts of greenhouse gases in the atmosphere, such as carbon dioxide and methane, trap some of this heat.
4. Increased greenhouse gases as a consequence of people's activities are trapping excess long-wave radiation and causing the atmosphere to overheat.
5. Some greenhouse gases are essential for humans to live on Earth. Without them our planet would be as cold as the moon's surface. But too much of

them may cause drastic changes in climate that could lead to severe human and ecological consequences.

■ **Some Causes of Global Warming**

- When fossil fuels such as coal, wood and oil are burnt they release carbon dioxide into the air. Carbon dioxide traps heat and prevents it from escaping into outer space.
- Increased industrialisation, intensive farming and vehicle usage are all using vast amounts of fossil fuels.
- A greatly increased use of fertiliser, huge herds of cattle and large areas of paddy rice are releasing extra methane gas.
- The use of chlorofluorocarbons (CFCs) in aerosols, fridges, foams and solvents causes about 14 per cent of global warming.
- America's population is producing 25 per cent of the global emissions of carbon dioxide.
- India, with 16 per cent of the world's population, is producing only 3 per cent of the carbon dioxide emissions.
- Deforestation is reducing the amounts of forest cover worldwide.
- Deforestation of these trees reduces the amount of carbon dioxide that can be absorbed from the air, and so adds to global warming.
- In 1973 the world's population was 3.5 billion. Today it is 6 billion people. This huge increase has led to a corresponding rise in the amount of fossil fuels that are burned.

■ **Consequences of Global Warming**

Negative effects

- Serious tropical diseases, such as malaria, could spread to temperate countries such as Ireland.
- There may be a higher risk of skin cancers among pale-skinned people such as the Irish.
- Trends in world tourism could be dramatically altered.
- Mediterranean regions may suffer a decline in tourism, as temperatures may be too high.
- Winter holiday resorts, for example ski resorts like Zermatt in Switzerland, could have much-reduced snowfalls, which would wipe out its tourist industry.
- If trends continue, predictions indicate that sea levels will rise by 0.2–1.4 m through thermal expansion and the melting of ice at the polar icecaps. Temperatures may rise by as much as 3°C over the next 100 years.

- This means that there would be greater extremes of weather, with freak storms and droughts more likely to occur (e.g. hurricanes in the southern United States).
- There would be disruption of agriculture, especially in marginal areas (e.g. the Sahel in Africa), causing local starvation and the mass extinction of plants and animals.
- Millions of people would be forced to migrate, as their lands would be swamped by the rising oceans, especially along delta areas in eastern Asia, where there are already extremely high concentrations of people (e.g. Bangladesh).
- In Europe the polderlands of the Netherlands would come under intense pressure from the sea, while much of the Wexford coast would be submerged, unless it is protected by dykes.

Positive effects for Ireland

- Increased global temperatures would increase Ireland's growing season and hence the range of crops that could be grown here. For example, grass production might increase by as much as 20 per cent and cereal production could also increase.
- Tourism could increase, as a consequence of a rise in both summer and winter temperatures.

Negative effects for Ireland

- Many low-lying coastal regions, such as the Wexford Slobs, could be flooded by rising sea levels.
- Increased heat could cause increased cloud cover and winter rainfall.
- The North Atlantic Drift might be deflected away from Ireland's shores, leading to a loss in fish species in Irish waters. In addition, winter temperatures could fall, rather than increasing, causing a freezing of our seas during winter.

■ Desertification in North Africa

Causes of desertification

- Desertification threatens the lives of hundreds of millions of people in sub-Saharan Africa.
- The Sahel region stretches for almost six thousand kilometres, east to west, across Africa.
- Until recent times 70 per cent of the Sudan was covered with tropical forest vegetation or savannah woodland. Many of the trees were cut down as the demand for cash crops increased to meet national debts.

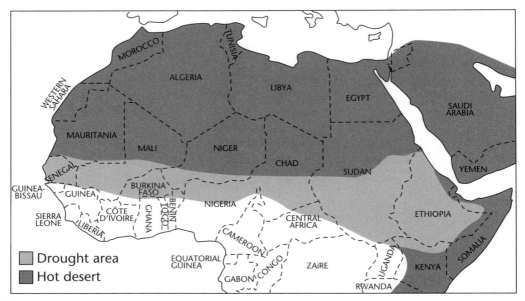

▲ Drought areas of North Africa

- Throughout the 1960s, countries in the Sahel became part of a global economy. They began producing goods such as cash crops for a global market.
- Because of this the area devoted to agricultural crops such as groundnuts increased, and encroached into marginal regions that were unsuited to such activities; farmers became dependent on cash crops for their livelihoods. When the rains failed, the soils were left exposed to the winds, which caused erosion of the soil.
- In colonial times, administrative centres grew up across the Sahel. This urban development continued into recent times, creating a huge demand for wood for fuel and construction.
- This deforestation, in association with global warming (the 'greenhouse effect'), has increased the water-supply problem in marginal areas and has led to desertification. Climate change is brought about by the interaction between the forest and the atmosphere.
- When trees are absent, so are clouds. With no clouds, the land becomes dried up by the equatorial sun. When rain does fall, the full force of the rain reaches the ground, and **sheet erosion** occurs. Nutrients and topsoil are washed into rivers and estuaries, creating problems of silting.

Consequences of desertification

- Crops have failed, farm animals have died and people have been left without adequate means of survival.
- Many children in the Sahel region have died of diseases associated with malnutrition, such as kwashiorkor.

- Many people have migrated southwards, causing overpopulation and further over-cropping and over-grazing of land.
- Many have migrated to cities, such as Niamey in Niger. This has increased the demand for wood as fuel, which in turn has led to more deforestation.

OPTION TOPIC 6
THE IMPACT OF SOCIAL AND POLITICAL DECISIONS

RURAL-TO-URBAN MIGRATION IN THE THIRD WORLD

- Rural unemployment caused by increased migration has led to urban migration in third world regions. Increased westernisation means agribusiness companies now control agricultural production on a global scale. Small landowners cannot compete and so they sell their farms.
- Rural-to-urban migration affects all countries, including Ireland. However, in third-world regions the rural populations are large and so migration to cities has happened, and is still happening, on a large scale.
- More than two-thirds of all the world's city dwellers now live in developing countries. Cities such as Mexico City, Sao Paulo and Calcutta have all added millions to their populations over the past 30 years.
- Thousands of people in the Sahel region of sub-Saharan Africa have migrated southwards as a consequence of desertification in the area. These people were farmers who tilled the land or grazed their cattle on their traditional homelands.
- Rural migrants have poured into cities out of desperation and hope, rather than having been drawn by jobs and opportunities.
- Because these migrations have mostly been composed of teenagers and young adults, an important additional impact has been exceptionally **high rates of natural population increase** (high birth rates with low death rates).

POLITICAL ACTIONS LEADING TO POLITICAL REFUGEES

■ Case Study: The Creation and Break-up of Yugoslavia

- The state of Yugoslavia was created in 1919 after World War One ended. The new state was created to deprive the Austro–Hungarian empire of its lands in the Balkans.
- The invasion of Yugoslavia by German forces in 1941 led to internal strife that caused the deaths of one-tenth of its population. Divisions created between ethnic groups led to lasting enmity after the war.

- Once the Communist system collapsed in 1990, these divisions arose once more.
- The ethnic cleansing of minority groups in the former state of Yugoslavia led to whole-scale migration of these minorities from some regions in the 1990s. Many migrants fled to neighbouring states such as Italy. About one thousand others came to Ireland as political refugees.

■ **Case Study: Northern Ireland**

- Northern Ireland was ruled by a Unionist majority government. Its state was formed in 1920 by the Government of Ireland Act which divided the island of Ireland into a Protestant-majority North and a Catholic-majority South.
- Since that time Northern Ireland had an almost two-thirds Unionist majority, until recently. Unionists were mostly Protestant and wished to be part of the United Kingdom.
- Nationalists were mostly Catholics who wished to be part of the independent Irish Republic.
- Unionists were culturally British and nationalists were culturally Irish.
- Political divisions spawned discrimination, repression, terrorism and periods of great violence.
- Violence led to migrations of Catholics from Protestant housing estates and Protestants from Catholic housing estates. This created ghetto communities in some areas.
- Severe violence initially occurred during civil-rights campaigns from 1969 to 1972. This led to migrations of some Catholics to the Republic, to new towns such as Shannon in Co. Clare where accommodation was available.

PATTERNS OF MIGRATION MOVEMENT

Rural-to-urban movements

- Most migrants are searching for employment, better education and health services.
- Many are fleeing ecological disasters, such as desertification in the Sahel.
- Millions flee from civil unrest such as civil wars, as in Sudan in 2004–2005.

Migration between third-world countries

- This accounts for 80 per cent of all international migrations.
- It results from a number of causes, such as employment seeking, ecological disasters, war or persecution.

Migration between third-world countries and the developed world

- This occurs from Central and South America to the USA.
- There is also migration from Africa, the Middle East and China to the EU and the USA.
- Poverty and lack of employment at home cause these migrations.

Migration from Eastern Europe to Western Europe

Up to 2 million people will have migrated between 2001 and 2006, because of:

- the sense of freedom created by the collapse of the Communist system in Eastern Europe
- poverty or poor living standards creating a search for a better standard of living
- the persecution of ethnic or cultural groups such as gypsies.

Migration from peripheral regions to core regions

This is fuelled by:

- a search for employment and higher living standards
- a search for better social and cultural centres, from education to nightlife.

Migration to Ireland since the 1990s

- Ireland was an area of out-migration, rather than in-migration, since famine times.
- The 'Celtic Tiger' economy created a need for many workers in all sectors.
- These include:
  - returning Irish emigrants who had worked abroad in the UK, EU or America
  - foreign nationals for work in the service industries
  - refugees, both political and economic, in search of sanctuary and a new life with hope for a better future.

■ Ireland's Refugees and their Status

- Ireland's attitude to refugees has changed since the 1990s. In the past they were a curiosity; now they are seen as a threat.
- Racial attacks are not uncommon.
- A 1999 survey found that 95 per cent of African asylum seekers suffered verbal abuse in Ireland.
- Twenty per cent of third-level students felt that illegal foreign nationals should be sent home.
- Many immigrants feel that the Irish State's attitude is racist and they are viewed as criminals. This view has been supported by racist comments from some elected government representatives.

■ **Positive and Negative Aspects of the 1996 Refugees Act**

Positive

- Refugees can't be sent home if they can prove their lives are in danger.
- Refugees can't be deported without being advised of their right to apply for asylum.
- Refugee status must be heard and processed fairly.
- A refused applicant has the right of appeal.
- Successful applicants have the same rights as native Irish to all work-related and social needs.

Negative

- Asylum seekers receive no free legal aid or rights to interpreters.
- They are not allowed to work or leave the country during the application process.
- They may have to wait years for their application to be processed.
- Economic migrants are not recognised as refugees.

OPTION TOPIC 7
INTERNATIONAL DEBT AND CYCLES OF POVERTY

CAUSES OF DEBT OF THIRD-WORLD COUNTRIES

- The rapid rise of oil prices in 1973 triggered a worldwide recession and caused debt repayments of third-world countries to escalate.
- Profits of oil-producing countries could not be reinvested due to world recession. So the banks offered huge loans to third-world countries as development aid. Many of these countries were ruled by dictators who invested unwisely in arms or useless development programmes.
- The United States in the 1980s raised their interest rates to attract overseas funds. This set in motion a spiral of increased rates throughout the developed world; so debts went out of control and poor countries could not repay even the interest.
- The IMF (International Monetary Fund) was set up by the banks of developed countries to restructure loans so third-world countries could repay their debts by doing the following:
  1. increasing cash crops
  2. reducing spending at home in services such as education
  3. stopping subsidising the price of foodstuffs, so essential foods became very expensive

4. introducing wage control to reduce inflation – but this made living even more difficult
5. devaluing national currencies to make exports cheaper, but that made imports more expensive
6. allowing the repatriation of profits of MNCs.

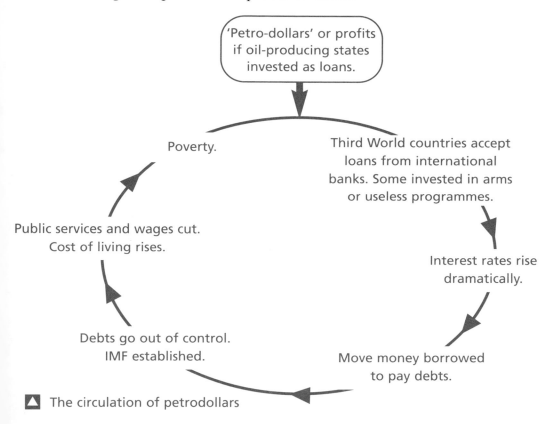

▲ The circulation of petrodollars

SOME SOLUTIONS TO INTERNATIONAL DEBT

- By 2000 the world's richest countries agreed to cancel about 30 per cent of total debt by third-world countries, and 100 per cent for some individual countries.
- Debt of the poorest African countries was cleared by the G8 countries in 2005.
- A large percentage of profits of multinational corporations in a third-world country should be reinvested in new industries or services in that country.
- Debt repayments should be reduced further to a level that allows these countries to develop their economy.
- Developing countries that have their debts reduced should be required to invest existing repayments to promote the development of self-reliance programmes throughout their country.

OPTION TOPIC 8
TYPES AND ADVANTAGES OF AID

WHO BENEFITS FROM AID?

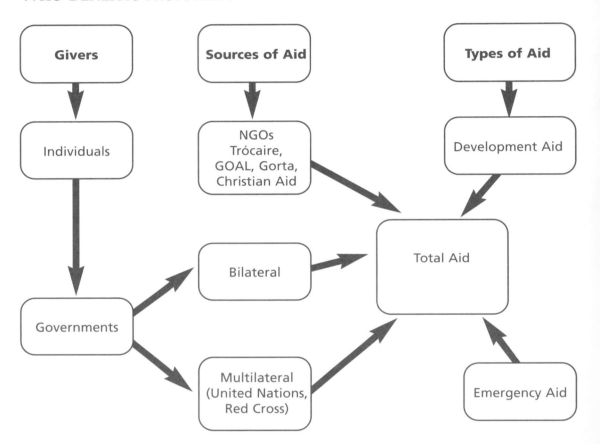

INTERNATIONAL AID TO THE SOUTH

■ **Aid Sources and Who Benefits**

1. **Non-governmental organisations (NGOs)**

   Voluntary organisations such as Trócaire, Concern and Gorta provide both emergency and development aid.

2. **Bilateral**

   This is direct aid from one government to another. Generally this type of aid is used to improve agriculture, education, health services, etc.

3. **Multilateral**

    International institutions such as the Red Cross, the United Nations and the World Bank provide both emergency and development aid.

■ **Emergency Aid**

Countless lives have been saved from famine and disease as a consequence of natural disasters such as earthquakes, famines and flooding by the provision of:
- food, clothing and shelter
- emergency personnel
- medical aid.

■ **Advantages of Emergency Aid**

- The supply of food, fresh water, medicines and shelter has saved countless lives.
- Modern transport systems have made the delivery of emergency aid much easier and more effective than in the past.
- Emergency aid does not upset local food producers in a negative way.

■ **Development Aid**

Development programmes in the LDCs (least-developed countries), such as Zambia, Ethiopia and Lesotho, for projects such as clean water supplies, farm livestock improvement, adult literacy.

■ **Advantages of Development Aid**

- Vital infrastructure such as water-supply pipes and wells, sanitation and new roads can give an initial boost to an emerging economy.
- Farm improvement schemes and education programmes help people to cater for their own future long-term needs.
- Health clinics develop skills of local people to cater for their communities' basic needs.
- Development Aid is called 'appropriate aid' because it serves the needs of local communities.

■ **Human Development**

- People who feel the need to help are given the opportunity to act as volunteers.
- APSO (Agency for Personal Services Overseas) recruits people to work on aid programmes.
- Development education provides workshops, seminars and resource packs to inform the Irish people and people of other countries of the dangers of xenophobia.

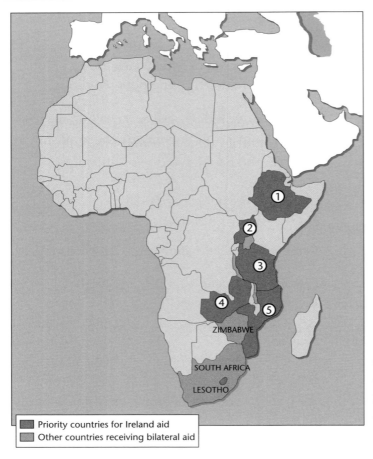

Priority countries for Ireland aid
Other countries receiving bilateral aid

◀ Sub-Saharan African countries that receive Irish and bilateral aid.
Activity: Identify the countries 1–5 that receive Irish bilateral aid.

▪ Disadvantages of Aid

- Tied aid may benefit the donor more than the receiver.
- Aid may cause the receiver to become dependent on the donor country, or if it is military aid it may be used by a government against an opponent who may be justified in their opposition to certain government programmes.
- Much tied aid is given to better-off third-world countries that may serve the political or economic needs of donor countries.
- Some powerful countries use it as a means of political control over weaker countries. For example the US government requested the use of Turkish airports as bases from which to bomb Iraq. When the Turkish government refused this request, the US immediately withdrew millions of dollars of aid from Turkey.
- Battles were fought in developing countries because governments and opposition groups were supported by the United States or Russian aid. Their **Cold War** was fought on others' soil.

- Aid has been used to encourage global free trade, which exposes the markets of poor countries to the products of rich countries.
- Loans may be given in the form of aid. Some of these loans are given on standard commercial terms, and may create crippling debt over long periods.
- The United Nations agreed that developed countries should set aside 0.7 per cent of their GNP as aid. Few countries have achieved this, with only Denmark, Norway, the Netherlands and Sweden having fulfilled the agreement.
- United States' aid, for example, is just 0.1 per cent of GNP.

THE ROLE OF NGOs

■ What NGOs are

- NGOs are non-governmental, private agencies that provide aid to developing countries. They include such agencies as Trócaire, Concern, Goal, Afri and Oxfam.
- They provide the following types of aid:
emergency relief, development aid, empowerment aid, and education awareness.

Advantages of NGOs

- Their independent status allows agencies such as Trócaire to work independently for justice in places such as South Africa, which was under Apartheid policies up to the 1980s.
- Because NGOs are relatively small organisations they do not get involved in mega-projects such as dam construction. They are mostly involved with community-based 'appropriate aid' projects that get local support.

Disadvantages of NGOs

- Competition between NGOs can lead to the agencies adopting 'starving-baby' images that may distort the First World's view of the Third World.
- The scale of funds at the disposal of NGOs is relatively small when compared to national government funding. Nevertheless leading Irish agencies' funds, Trócaire and Concern, totalled over €26 million in 2000.
- Some NGOs use 'child sponsorship' schemes where individual children in the Third World are sponsored by individuals in the First World. However, this could make such third-world children feel personally in debt to these sponsors and so may reduce their sense of individuality.

OPTION TOPIC 9

LAND OWNERSHIP PATTERNS AND THEIR IMPACT ON DEVELOPMENT IN NINETEENTH-CENTURY IRELAND

If the people who farm own the land that they work, then they will have a much better opportunity to develop both economically and as humans.

NINETEENTH-CENTURY IRELAND

- Most of the land was owned by a small number of landlords.
- The land was worked by a large number of tenant farmers.
- Landlords demanded high rents called 'rack rents'. This made Irish tenant farmers poor.
- Most landlords were absentee landlords.
- The better a tenant farmed his land, the higher his rent.
- Tenants held land by verbal agreements on a year-to-year basis, so they could be evicted at the time for rent review each year.
- In 1849 over 13,000 families were evicted from their holdings.
- Many landlords were fair to their tenants, but some were ruthless to them.
- Landlord John Adare of Derryveagh in Donegal evicted sixty families in the middle of winter.
- In 1879 Michael Davitt, supported by the MP Charles Stewart Parnell, set up the National Land League.
- The struggle over the future ownership of the land led to the 'Land War'. That led in turn to the Land Acts, which eventually transferred ownership of the land from the landlords to the tenants through tenant purchase schemes.
- The Ashbourne Act and the Balfour and Wyndham Act eventually led to its being compulsory for the landlords to sell their land to their former tenants.
- Land ownership now gave some hope for the new farm owners to be able to improve their livelihood.
- However, in the short term they had only a poor local market.

■ Irish Co-operatives

- Sir Horace Plunkett founded the Irish Co-operative Movement and in 1889 he started his first co-operative in Doneraile in Co. Cork.
- By 1914 there were over 1,000 co-operatives. Since then many of them have amalgamated and grown into huge businesses. One of these, the Kerry Group, is a multinational business.

■ **Local Enterprise Boards**

The Local Enterprise Boards were established in the 1990s with encouragement from the government.

The LEBs usually take the form of private limited companies managed by local interest groups, such as community councils, chambers of commerce and farmers.

Many LEBs do the following:

- manage the Leader Project, which is an EU-funded rural-development initiative. Leader funds up to 50 per cent of capital investment and up to 100 per cent of training costs for local development projects.
- administer nationally funded Local Development Programmes. These programmes are used to train people like travellers or long-term unemployed people for the workplace. They give advice on things like CVs and interview techniques.

OPTION TOPIC 10
HUMAN EXPLOITATION

■ **World Trade and Global Exploitation**

- It is believed by some people that third-world poverty has increased as a result of global trading.
- The richest 20 per cent of the world's people control 84 per cent of global trade.
- The poorest 20 per cent control less than 1 per cent of global trade.
- Multinational corporations such as Volkswagen have turnovers of twice the GNP of Bangladesh.
- The turnover of Nestlé Corporation is 20 times greater than the entire GNP of Nicaragua.

REASONS FOR LACK OF CONTROL OF GLOBAL TRADING BY THIRD-WORLD COUNTRIES

■ **Reliance on a Single Commodity or Raw Material**

- Third-world countries were colonies that supplied commodities such as coffee, tea and cotton for factories in rich countries.
- Prices for these commodities from developed countries are controlled by powerful multinational corporations, such as Chiquita and Nestlé.

- Some third-world countries are totally reliant on a single commodity for export, so if the price for that commodity falls it is a disaster for that country.

■ Unfair Trading

- The prices of goods from developed countries to third-world countries have risen hugely, but the prices of goods from third-world countries to developed ones have fallen.
- So they have had to sell more and more to buy the same amount of goods.
- In 1972 Uganda sold 6 tons of cotton to buy one truck.
- Today it must sell 35 tons of cotton to buy a similar truck.

■ Fluctuating Prices

- Because prices of commodities vary greatly from year to year, third-world countries are unable to make long-term plans for development. Sometimes this is due to poor advice from the World Trade Organisation (WTO).
- As more commodities are grown for sale to pay debts, there can be a glut of some commodities in a particular year. This causes a fall in prices.

Case study: Coffee

- Coffee is the world's second most important commodity after oil.
- Some countries, such as Burundi and Ethiopia, depend on coffee for most of their income. A bad harvest or a sudden drop in price can bring bankruptcy to such countries. For example in 1989 the price of coffee fell by one-third in a single week.
- The price of coffee was regulated in the past. But rich, coffee-consuming countries and the World Trade Organisation ended this practice.

■ Cheap Labour

- Many manufactured products, such as clothing and footwear, are now produced in third-world countries. New technology and multinationals (or TNC, transnational corporations) have been responsible for this trend, creating what is called 'globalisation'.
- New technology has reduced the need for highly paid, skilled labour.
- Multinationals have brought welcome work for third-world countries, but they gain their profits from the poorly paid and badly treated workers.
- Most of the profits of TNCs return to where they have their headquarters.
- For example, the legal minimum wage in Indonesia is $1.27 per hour: but 12,000 factories pay less than 40 per cent of the minimum wage.
- Female workers make up 80 per cent of the workforce.
- A normal working week is 50 hours, with no payment for overtime.

■ **Exploitation at Local Level**

- Many women are exploited by being made a part of illegal trafficking to the EU and Ireland.
- These women come from Eastern Europe, the Middle East and Africa.
- They begin their journey as economic migrants.
- Members of Mafia-style gangs promise them guaranteed work to coax them into travelling. They smuggle the women into Europe, and then they must live in total submission to their captors, as they have no passports or work visas.
- They are frightened, and are without the local language, so they become slaves.
- About one-third of the slave women become prostitutes, especially in the Netherlands and Belgium.
- Another one-third work as maids for miserly sums. The rest work in sweatshops for little pay.

CASE STUDY: BOYS AT LETTERFRACK IN CO. GALWAY

- Up to 200 boys, aged six to sixteen and from poor families, lived at any one time as inmates of an industrial school.
- It was a reform school, where they were to learn the skills necessary to become productive and responsible members of society.
- From the outside it looked ideal and it boasted of services provided in tailoring, boot-making, carpentry, baking, cart-making and other services.
- However, on the inside things were quite different.
- Boys as young as six worked long hours for no pay. It was forced labour, slave labour.
- Physical and sexual abuse were common occurrences.
- It was closed in 1974.

OPTION TOPIC 11
GENDER ROLES

Tradition, injustice and poverty are three factors that have seriously affected the role of women in society.

- Certain societies, such as Muslim ones, have allowed daughters to inherit only half as much as sons.
- Women could not file for divorce against their husbands.

- The Taliban regime in Afghanistan forced women to cover their entire bodies, even their faces. They could not attend schools or work outside the home.
- China's 'one child per family' policy led to the deaths of many infant girls, as there was a preference for sons to carry on the family name. It also led to selective abortion of many unborn baby girls.
- In many poor countries boys are given preference over girls for facilities such as education.
- Girls can be forced into arranged marriages with distant cousins many times their own age.
- Women must play many roles, as wives, mothers and subsistence farmers, sowing and reaping the crops.
- Women in transnational corporations must work for less pay than men.

WAYS OF CHANGING GENDER ROLES

◼ Self-help

- Organised groups of women can challenge their position, assert their social rights and improve their income.

Case study: Kassassi in Sierra Leone

A group of women organised the drainage of 100 acres of swampland, then sold their crops locally during the off-season when other crops would not grow. This increased their income and that of local children, who worked to save money for their education.

◼ How to Break the Cycle of Poverty

Education

- Lack of education creates cycles of poverty.
- Improved school facilities can include necessary skills learned in apprentice-type programmes.
- Adult education programmes can help mothers to learn and pass on their new knowledge and desire for learning to their daughters.
- Social education for men may help them to encourage women into education and economic services

Workplace reform

- Strong trade unions are needed to work on behalf of women's rights in places where they are exploited.
- Laws need to be introduced that force equal pay for equal work. In Ireland, the Employment Equality Act was introduced in 1997.
- Additional facilities, such as crèches, and maternity leave encourage mothers to work outside the home.

Aid for women

- Better-designed aid packages would help to facilitate the economic empowerment of women.
- Banking structures should be rethought so that poor women can take out loans to invest in small economic projects. The Grameen Bank in Bangladesh loans money to poor women to buy seeds, cattle, farm machines and even land.

OPTION TOPIC 12
THE SUSTAINABLE USE OF RESOURCES

Sustainable use of resources means meeting the needs of the present without endangering the requirements of future generations.

THE KYOTO PROTOCOL

The Kyoto conference held in Japan in 1997 set a target of an overall reduction of 5 per cent in the production of greenhouse gases by 2012. This was agreed by all participating countries, including all those of the developed world.

However, since then:

- The USA and Japan demanded the right to buy 'pollution quotas'.
- In 2001 The United States abandoned its Kyoto commitments.
- Ireland agreed to keep its 2012 emissions to 13 per cent above its 1990 levels. But by 2000 Ireland's emissions had exceeded 20 per cent. However, Ireland intends to do the following:
  - Coal burning at Moneypoint is to be phased out or changed to natural gas supply.
  - More energy is to be sourced from wind power.
  - There is to be a reduction in carbon emissions from vehicles, with higher taxes on fuel to reduce consumption.
  - Energy-efficiency certificates are to be introduced for houses for sale that were built before 1991.
  - A grant-based tree-planting programme is to be initiated to reduce carbon emissions.

THE SUSTAINABLE USE OF FORESTS

Sustainable use of forests means:

- the conservation of some forest regions
- the managed use of other forests
- afforestation programmes.

■ Forestry in Scandinavia

- Over 200,000 hectares of forest are harvested each year.
- 70 per cent is reforested manually. Natural regeneration occurs in the remaining area.
- Disease in forests is carefully controlled.
- Improved land drainage helps the faster growth of trees.
- Careful use of fertiliser increases yields.

■ Sustainable Forest Use in Cameroon

Case study: Korup project of Cameroon

- This involves the integration of a forestry programme and a farming programme.
- A national park has been created, where all trees are permanently protected.
- A buffer zone exists between the national park and the intensive farming region surrounding it, where sustainable farming of small farm units can be carried out.
- Agriforestry has been introduced, involving the rotation of food crops with trees.

■ The Sustainable Use of Fish Stocks

Why were fish stocks reduced?

- improved technology such as sonar to locate fish shoals
- large trawlers that could stay at sea for many weeks
- monofilament nets that fish could not see.

Activity

Look at the graphs on page 233.
(i)   Identify the trend in the spawning stock of Irish Sea cod from 1968 to 2002.
(ii) State the year that spawning stock fell below its precautionary level.
(iii) Identify and explain the differences in the trends in 1968–1982 and 1982–2002.

By 2000, two-thirds of fish stocks in EU waters were approaching commercial extinction.

Conservation measures

- Quotas were introduced on the quantities of fish that each state could catch. This was called TAC: total allowable catch.
- Fishery exclusion zones limit fishing in rich, shallow water near coasts.

- Waters within 10 km of Irish coasts could be fished only by Irish trawlers.
- The EU proposed to reduce the size of its fishing fleet by 40 per cent.
- Greater policing of coastal waters by fishery protection vessels was introduced.

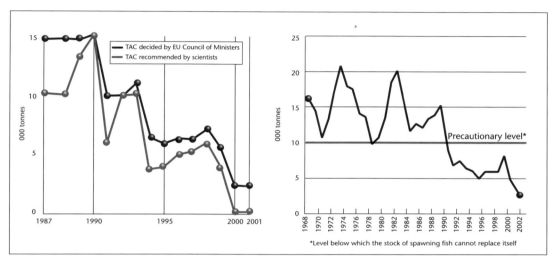

▲ Total spawning stock of Irish Sea cod

OPTION TOPIC 13
FAIR TRADE

Globalised trade is driven by the aim of achieving constantly greater profits for multinational corporations.

This policy increases inequalities between rich and poor regions.

Globalised trade is not aimed at improving levels of human development.

THE MEANING OF FAIR TRADE

- To enable all producers to earn an adequate living by prices for produce, such as coffee, being index-linked to the prices charged by developed countries for manufactured goods such as coffee products.
- That fair wages and safe working environments be provided for workers.
- That goods sold can be economically and ecologically sustainable.
- That third-world countries gain a reasonable degree of control over their own country's economy.
- That small-scale producers be able to contribute to a country's exports.

■ **The Unfair Existing Global Trading System**

- The prices of third-world commodities fluctuate widely on the world market. This policy is encouraged by wealthy nations to their own benefit.
- The products of most multinationals are made in poor countries by people who work in poor working environments, such as sweat shops, for minimum wages.
- Some products, such as tropical hardwoods, are produced by destroying tropical forests.
- Trade barriers are created against some poor countries for political or other reasons.
- Multinationals (transnational corporations) have increased their control of world trade. Seventy per cent of global trade in wheat is now controlled by six multinationals.
- Changes by the International Monetary Fund have reduced the control of governments over the activities of multinational companies.

■ **What does the 'Fairtrade Mark Ireland' mean?**

- The Fairtrade Mark Ireland and other Fairtrade labelling organisations have been established to promote fairer trading practices of quality products.
- The Fairtrade Mark is a guarantee to consumers that the producers have been paid a fair price for their produce and that the producers work in safe and decent working conditions.
- Some EU countries use a different label. These include the **Max Havelaar** label in the Netherlands, Belgium, France and Switzerland. The **Transfer** label is used in Germany, Austria and Luxembourg.

■ **How does the Fairtrade System work in Practice for Nicaraguan Coffee Producers?**

- The first-world importers deal directly with the producers to eliminate middlemen.
- The producers are offered a guaranteed minimum price that is higher than the international market price.
- Some money is paid in advance to the producers and an additional premium is given for some local community project.
- The producers must guarantee to provide a safe working environment and an eco-friendly production system.
- The Fairtrade Labelling Organisation International (FLO) certifies that the production process of the coffee has met its standards.

- The consumers are willing to pay about two cents extra per cup for this Fairtrade product.
- Sales by Fairtrade labels account for about €200 million annually.

Has Ireland responded positively to Fairtrade produce?
- Since March 2001 three major supermarket chains in the Republic of Ireland stock Fairtrade bananas that come from Ghana and Costa Rica.
- Some small family-run shops sell Fairtrade goods.

■ How has Fairtrade Affected some Producers?
- In the Volta Rivers Estate in Ghana workers are now paid twice the national minimum wage and own 25 per cent of the company's shares.
- Toilets and showers are installed on the estate and water is piped to the workers' village.

■ How does Fairtrade Produce Affect the Consumer?
- Healthier foods are produced by Fairtrade systems.
- A more eco-friendly production system uses fewer chemicals than other growers.
- Banana crops are sprayed only 10 times annually, compared to 40 sprayings on other plantations.
- Chemicals such as paraquat are no longer used on Fairtrade crops. Some 20 per cent of male banana workers were left sterile in recent years in Costa Rica after handling toxic chemicals.

CASE STUDY: THE BARCELLONA CO-OPERATIVE IN THE PHILIPPINES
- Barcellona is a small region on the island of Luzon.
- The local people were skilled in craft industries such as basket making.
- A British fair-trade organisation, called Traidcraft, helped in the setting up of this co-operative by providing training in dyeing, colour selection, design and pricing.
- Traidcraft agreed a fair price for the goods and paid half of the total agreed price for their first order.
- This prepayment helped the people of Barcellona set up a small production unit of 12 members.
- The aims of the co-operative included the provision of basic needs for the local community.
- Woodwork and metalcraft activities were organised.

- Forty per cent of the co-op's profits were set aside for a child-care centre.
- Training courses in techniques of herbal healing were set up, as there was only one doctor available for the 14,000 local people.
- Essential herbs were grown to create an adequate supply of medicines.
- Traidcraft helped the poor local people of Barcellona to meet their own basic needs and to take control of their own lives.

OPTION TOPIC 14

JUSTICE AND MINORITY GROUPS

'The level of development of any society can be measured by the degree of justice with which that society treats its weakest members.'

TRAVELLERS IN IRELAND

- There are about 22,550 members of the Traveller community.
- Travellers are native Irish people with an identity, culture and tradition distinct from the majority of people in Ireland.
- Travellers are nomads by tradition and follow certain patterns for social, cultural and family events. Many use 'Halting Sites' where they live throughout the winter months so their children can attend school. Fine summer weather is used to travel from place to place trading.
- Familiar roadside sites on byroads are used for setting up camp.
- Strong family ties and religion are important to Travellers. Loyalty to the extended family plays an important role in Traveller society.
- Music is important to Travellers and many are highly talented and skilled musicians.
- Travellers' language is called 'Cant', and is still spoken.

■ Irish Injustice

Travellers were recognised in 1991 by a European Committee of Inquiry as 'the single most discriminated-against ethnic group' within the European Union.

Racism

- 42 per cent of Irish people hold negative attitudes towards Travellers.
- 59 per cent of Irish people would not welcome a Traveller as a next-door neighbour.
- 93 per cent would not accept a Traveller as a part of their family.
- 71 per cent of pubs discriminate against Travellers.

Recent legislation

- The 1991 Prohibition of Incitement to Hatred Act forbids incitement to hatred of Travellers.
- The 2000 Equal Status Legislation outlaws discrimination in the provision of all services to Travellers.

Education

- The segregation of Traveller children from other children in some primary schools prevented a sharing of their traditions, leading to distrust between communities.
- The education system was used as a way of diluting the cultural traditions of Traveller children.
- Only about 1,000 Traveller children attend secondary school at any one time.
- Only 100 Traveller children continue to fifth year in secondary school. Many leave school at fifteen years of age.
- Lack of education widens the social and prosperity gap between Travellers and settled people.
- The 1995 White Paper called for fully integrated classes that respected Traveller culture. This is known as intercultural education.
- Some teachers who focus on special needs visit halting sites to encourage Traveller children to remain at school.

Accommodation and living conditions

- In 1998, 24 per cent of Travellers still lived in unserviced roadside sites.
- Roadside sites have none of the basic services such as running water or toilet facilities.
- Untidy roadside sites create tension between settled local people and Travellers.
- Only 2 per cent of Travellers live to sixty-five years of age or more.
- The living conditions of Travellers make success in school almost impossible.
- The 1988 Housing Act gives state recognition of the need for Traveller-specific accommodation. This should take into account that extended families often live together and that Traveller families tend to be larger than the national average.
- A comprehensive range of options should include well-serviced halting sites, group housing schemes and conventional housing.

Health

- Infant mortality among Travellers is three times higher than among the settled community.
- Life expectancy of Travellers is ten years less than that of settled people.

Ways to improve the health of Travellers

- If the health of Travellers is to improve, their living conditions must also improve.
- National medical cards must be made available, so that treatment may be carried out in all health regions.
- Medical liaison officers for advice on treatments and prevention of sickness would help.

■ Bonded Labour in Pakistan

- Bonded labour is the most common form of slavery in the world today.
- About 20 million people are in bonded labour.
- About 4 million people are in bonded labour in Pakistan.
- Bonded labour exists in several third-world countries.
- Bonded labour develops when a person borrows money from an unscrupulous moneylender.
- In return the borrower agrees (bonds) to work for the moneylender until the loan is paid.
- Wages paid by the moneylender are low and insufficient to meet the interest, so the loan continues to increase.
- The bonded labourer/borrower is persuaded/forced to work indefinitely and without wages for the borrower.
- Other family members may be forced to work against the loan and this can continue for generations, so that some family members are born into slavery.

In 1991 the Pakistani government banned the practising of bonded labour. The Lahore High Court has outlawed the loaning of money in return for bonded labour.

But:
- Moneylending landlords are people of influence and few are charged with the crime.
- The Pakistani government denies that bonded labour is common in Pakistan. They state that there are only 7,000 people in bonded labour. Human-rights groups claim that over 4 million Pakistani people are in bonded labour.
- There are no work-related government schemes for people who are freed from bonded labour.

■ **The Campaign for an End to Bonded Labour**

- High Court rulings have helped free over 12,000 bonded labourers.
- The Rugmark label is an indication that no illegal child labour was involved in manufacturing.
- Voluntary organisations such as the Bonded Labour Liberation Front of Pakistan constantly seek legal means to fight slavery.
- The Irish NGO Trócaire promoted a Lenten campaign to oppose slavery.
- It supported local groups, such as NCJP, to fight slavery.
- It paid for education, training and grants to help freed bonded labourers.
- It encouraged the Irish government to use its influence to highlight the injustice of bonded labour.

OPTION TOPIC 15
SELF-HELP DEVELOPMENT IN THIRD-WORLD AND PERIPHERAL REGIONS

- Globalised trade has concentrated increasing wealth and power in the hands of just a few individuals and corporations.
- It is increasingly important that people-centred development is undertaken by self-help development schemes in which responsibility rests with members of each scheme.
- Such groups can link up with and gain financial and other support from larger, well-established development bodies. These include:
  - NGOs such as Trócaire, Oxfam and Christian Aid
  - women's movements that work for basic salaries and abolition of discrimination against women
  - cultural movements that help strengthen local cultures to counteract influences from Westernised globalisation.

CASE STUDY: SELF-HELP HOUSES IN LUSAKA, ZAMBIA

Third-world countries cannot afford the provision of social housing, so they encourage communities, with the help of the World Bank, to organise the building of their own homes

■ **How Does this System Work?**

- Groups of about 20 to 30 local people band together to organise the building project.

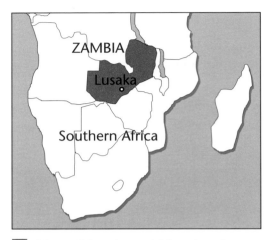

▲ Map of Southern Africa showing location of Lusaka, Zambia

- The World Bank and the government provide:
  - a suitable site of about 10 hectares with a water supply on site
  - enough money to buy materials for simple homes.
- The group provides the labour and if they do the work for nothing they will use the saved money to add electricity and road surfacing.
- A school or clinic may be added at a later date.

This scheme created self-esteem and community spirit among the local people.

CASE STUDY: SELF-RELIANCE IN KERALA, INDIA

Regional government initiatives in Kerala have achieved the following:

- The people of Kerala in southern India live on average more than 10 years longer than other Indians.
- They enjoy better health, education and transport services.
- There are fewer inequalities between male and females and between castes.

These results have been achieved in the following ways:

Land reform and financial security

- Over 1 million people were given direct access to land.
- Local co-operative credit was arranged to buy livestock and inexpensive hand tools.
- Community-based schemes provided clean, fresh water and primary healthcare to reduce disease and death rates.
- As people became more financially secure they had fewer children. Local family-planning campaigns helped reduce birth rates.

CASE STUDY: SELF-HELP ENVIRONMENTAL INITIATIVE IN THE IRISH HOME

To improve our environment we could do the following:

- Recycle our waste. More than half the waste we produce could be recycled.
- Recycling reduces energy costs.
- Recycling bins are located in almost all communities.
- Waste collection from January 2005 will be charged for by weight.
- This will encourage people to separate their waste for recycling purposes.

CASE STUDY: SELF-HELP IN KILTIMAGH, CO. MAYO

- There was a severe depression in Ireland during the 1980s.
- By 1988 over 75 per cent of the 17-to-25 age group of Kiltimagh had migrated from the region.
- The rural region of Kiltimagh was composed of small farms with poor land.
- Forty per cent of the buildings in the town were derelict.

■ What did the Local Community do to Overcome its Difficulties?

- The Kiltimagh Integrated Rural Development (IRD) was formed in 1989 to halt economic decline.
- A head office with a full-time manager and staff was established.
- The physical restoration of the town was pursued.
- Cables were put underground; the restoration of the square and the redesign of street lighting in a traditional design were carried out.
- Shop fronts and signs were re-created in keeping with traditional nineteenth-century patterns.
- Tourist potential was identified and angling was promoted in the region.
- A camper site was created and traditional annual festivals were set up to draw tourists to the region.
- Cultural and artistic developments were promoted by an arts manager.
- State agencies and local government help with these activities.

Results

- There has been a 6 per cent rise in local national schools.
- The town has been chosen as the location for the National Centre for Rural Development.
- Local businesses have increased their turnover.
- 350 more people are at work now in Kiltimagh than before the redevelopment.
- The population of Kiltimagh rose by 200 between 1996 and 2002.

SELF-HELP PROJECT IN EASTERN ETHIOPIA: THE EASTERN HIGHLANDS PROJECT

Aims:

- to increase foodstuffs and farm income in a sustainable way
- to improve basic services
- to restore local natural resources.

The project has a number of activities

- **Crop production**: farmers are provided with plentiful supplies of seeds and trained in crop-production techniques.
- **Livestock production**: Farmers are assisted in animal breeding, poultry rearing, bee-keeping and the production of animal feed.
- **Soil and water conservation and afforestation**: Seed nurseries are established and trees and grasses planted in gullies to reduce erosion.
- **Education**: Primary schools are being established in the area. Adult classes are conducted in environmental awareness, HIV awareness and family planning.
- **Public health**: Self Help supports the staffing and training of clinical staff, who focus on community vaccination programmes, health education and child nutrition.
- **Women's programme**: Self Help helps women to generate an income by learning skills such as needlework, dressmaking and market gardening.
- **Water supply**: Wells are bored for small irrigation systems and domestic water supplies.

OPTION 2:
CULTURE AND IDENTITY

OPTION TOPIC 16
RACIAL AND ETHNIC GROUPS

- A person's race or racial group cannot be changed.
- Race is a biological inheritance.
- Race refers to physical characteristics such as skin colour, height, hair type, physique and shape of head.
- These characteristics are passed through genes from parents to offspring.

ETHNIC GROUPS

- 'Ethnic' refers to a minority group with a collective self-identity within a larger host population, such as Italians or Irish in New York, or Chinese in Ireland.
- The Aborigines of Australia who were cut off on the once-isolated island continent are an ethnic group.
- The Kalahari Bush people, who were isolated in the Kalahari semi-desert region of southern Africa, form another such group.

■ Skin Colour and Race

- In the past racial groups were classified by skin colour.
- They were divided into five groups: white, yellow, red, brown and black.
- Scientists today believe that dark- and light-coloured skins developed because of humans adapting to their environments.
- Skin colour is a result of the presence of melanin-producing cells in our bodies.
- We all have the same number of melanin-producing cells.
- In dark-skinned people these cells produce 43 times more melanin per cell than in light-skinned people.
- Melanin is necessary to combat the effects of ultra-violet light by absorbing dangerous rays, and so it protects us against cancer.
- The greater the amount of sunlight, the greater the need for melanin to protect against skin cancer.

- Dark skin also was needed to protect against the effects of strong sunlight on the production of folic acid.
- Groups farther away from the Equator needed some ultraviolet light to create vitamin E, so humans developed genes for creating light-coloured skin.
- Therefore different skin colours developed at different latitudes, so that people could live healthy lives in those places.

Race and genetic make-up

- 75 per cent of all the genes in people are identical. Of the remaining 25 per cent, about 85 per cent would be similar if people belonged to a racial grouping, such as Norwegians, a sub-group of Caucasian.
- Only 6 per cent would be different if people were from what we call separate races.
- So any person's race accounts for only 1.5 per cent (or 6 per cent of 25 per cent) of his genetic make-up.

There are 5 recognised ethnic or racial groups:

- **Caucasians**
  Europeans and people of European ancestry, brown-skinned people, such as Arabs and people of the Indian subcontinent
- **Northern, Central and East Asians**
  Chinese, Inuit, Samis and American Indians (Amerindians)
- **Africans and black people of African descent**
  (such as African-Americans)
- **Black Australian Aborigines**
- **The Bush people of the Kalahari.**

OPTION TOPIC 17
THE IMPACT OF EUROPE ON WORLD MIGRATION AND RACIAL PATTERNS
THE COLONISATION OF THE NEW WORLD

- Emigration to the Americas was a 'release valve' for many overpopulated European regions. One-third of the population growth in the USA between 1850 and 1910 was due to people of working age emigrating from Europe.
- The European-settled lands of North America, Australia and New Zealand, parts of South America and South Africa provided:

1. cheap food supplies for Europe's growing population
2. raw materials for its industries.

- The effects of European colonisation on the Native American people were catastrophic.
- Over 24 million native Americans in Mexico died between 1519 and 1600.
- The native people of what is now the USA declined from 5 million in 1500 to 60,000 in 1800.
- These native people were devastated by epidemics of infectious diseases such as smallpox, measles, influenza, tuberculosis and chickenpox.
- Over 12 million Africans were transported to the Americas as slaves.
- Chinese and Indians were attracted to South-East Asia to work in tin mines and rubber plantations. Chinese form 28 per cent and Indians form 8 per cent of Malaysia's population.
- Post-colonial unrest, such as civil wars and military coups, have devastated African states and slowed their economic progress.

■ **Multiracial Societies**

Case study: Britain

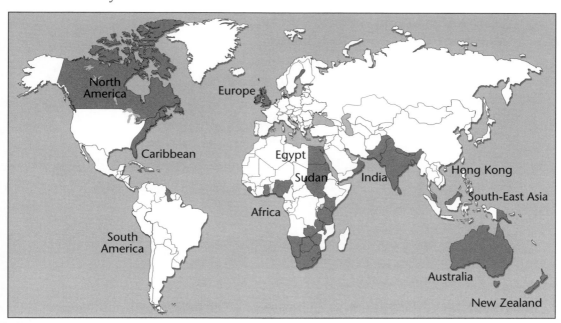

▲ The former colonies of the British Empire

- The ethnic or multiracial population in Great Britain has evolved from former British colonies in the Caribbean and South Asia during the 1950s and 1960s.

- Ethnic minorities make up 8.8 per cent of the population of Great Britain, and accounted for 7.2 per cent of all people of working age in 2000.
- People from ethnic minorities remain concentrated in the larger urban areas, especially in Greater London, and in some cases form a majority of the local population in certain districts, such as Notting Hill.
- At the peak of immigration in 1961, some 50,000 people arrived from the West Indies in one year.
- This migration was encouraged by the offer of employment from London Transport and the National Health Service.
- Britain used to be the centre of the 'triangular traffic', whereby British ships took goods to Africa, and exchanged them for slaves. Then the same British ships transported the slaves to the Caribbean and North America before returning home with industrial raw materials such as cotton.

Case study: France

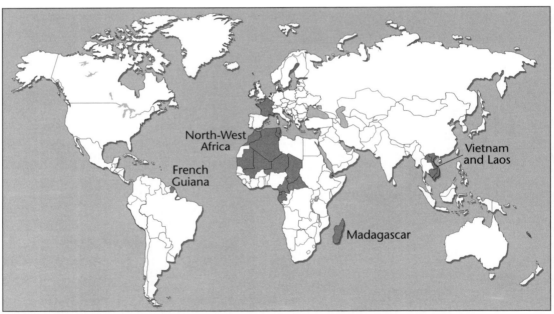

▲ This map shows the extent of the former French Empire

- Like Great Britain, France is a former colonial power.
- About 14 million French citizens, nearly one-quarter of the total population, have at least an immigrant parent or grandparent.
- A large share of the post-war immigrants and their offspring come from former French colonies in North Africa, sub-Saharan Africa, and from countries in South-East Asia.

- Immigrants also came from colonies in the Caribbean.
- This population is concentrated in the suburbs and urban centres such as Marseilles and Lyon.
- Most immigrants live in ghetto-like communities.
- Large past migrations and high birth rates among immigrants have made Islam the second-largest religion in France.
- Many immigrants, especially those from Algeria and Morocco, came to France as 'guest workers' in the 1960s and 1970s.
- Migrant workers performed jobs not wanted by the French, because the jobs were poorly paid, unpleasant, dirty or involved heavy manual work.
- In France, all citizens are deemed equal and indistinguishable in relation to the state. Be their origins Algerian, Senegalese or Corsican, French citizens are deemed identical in their 'Frenchness'.

■ **Racial Mixing**

Case study: Brazil
- Brazil has a population of 167 million people. About 80 million of them are people of black African descent.
- These are descendants of the slaves that were brought to Brazil by the Portuguese to work their sugar and cotton plantations along the north-east coast.
- Of the 80 million black people, about 67 million of them have combined European, African and Amerindian ancestries.
- The remainder, forming 55 per cent of the population, are mainly of European origin, the descendants of immigrants from Portugal, Italy, Germany and Eastern Europe.
- The Japanese population numbers about 1.3 million.
- Ethnic mixing is so great that hardly any group is unaffected.
- Only about 275,000 Amerindians survive. In the hunger for land and minerals, many have been forced from their forest homelands.
- Few black people are involved in politics and even fewer have positions of power within government, even though they form almost half of Brazil's population.

Case study: The United States of America
- America is a multiracial society where racial mixing has been slow to occur.
- The African-American group is the only group to have involuntarily emigrated to the USA.

- Only 44 per cent of blacks agree that race relations in the USA will eventually improve.
- Racial mixing has increased, especially among the young population. For example, 70 per cent of Italians born after 1970 have mixed ancestry from outside their ethnic group.
- Roughly 99 per cent of African-American women and 97 per cent of African-American men marry within their ethnic group.
- Many coloured people live in ghetto communities; and ghettos have come to symbolise the place of the poor within cities.
- Some people believe that with no work, no income and no property the only way to achieve a masculine identity is through crime and gang membership.

OPTION TOPIC 18
RACIAL CONFLICT

RACIAL CONFLICT IN THE UNITED STATES

- The African-American group is the only ethnic group that emigrated to the United States involuntarily.
- Slavery was introduced into the southern states to create cheap labour for cotton and tobacco planters.
- At the end of World War Two, Americans showed increasing concern over racial discrimination.
- In the 1950s the emergence of the Civil Rights Movement resulted in a revival of Ku Klux Klan organisations.
- The most important of these was the White Knights of the Ku Klux Klan, led by Robert Shelton.
- In the South, especially in the states of Alabama and Mississippi, violence against blacks by whites was rarely fully investigated.
- Lynching was still used as a method of terrorising the local black population.
- A civil court action by a mother against the Klan over the lynching of her son was successful. She was awarded $7 million and the Klan had to sell all its assets to pay the fine.
- Most civil-rights demonstrations stressed non-violence. But the demonstrations sometimes caused tension that resulted in violence.
- Martin Luther King, the black leader of the Civil Rights Movement, who was a Nobel Peace Prize winner, was assassinated.

Racial Conflict in India

Racial conflict in India may be looked at under two headings:

- the caste system in India
- northern and southern Indians.

■ The Caste System

The caste system usually refers to the groups of society into which the people of India are divided by religious laws.

- In general it means a **hereditary division** of any society into classes **on the basis of occupation, colour, wealth or religion**.
- India has four castes, the highest-ranking group being the Brahmans. **All others, who do not belong** to any of these four groups, become outcasts or **untouchables.**
- With the **introduction** of **British systems** to India, **the castes became rigid social divisions.**
- No one could rise to a higher caste than the one into which they were born.
- India today has become more flexible in the customs of its caste system. Urban people are less strict about the system than rural people.
- In cities, different castes of people intermarry and mingle with each other.
- In rural areas there is still **discrimination based on castes** and on some people being **untouchables**.
- Most of the degrading jobs are still done by the Dalits or untouchables, while the Brahmans remain at the top of the hierarchy by being doctors, engineers and lawyers.

■ Northern and Southern Indians

- The people of India belong to **all the major racial groups.** However, **Caucasians make up 90 per cent of the population.**
- People of Asian descent live mainly in the Himalayan Mountains, the highlands of the north-east and central India.
- The British promoted religious, ethnic and cultural divisions among their colonists to keep them under their control.
- The British promoted the idea that India is a land of two races – the lighter skinned Aryans, speaking a language of European origin, in the northern half of the country; and the darker-skinned Dravidians, who speak a different language, in the southern half.
- European thinkers of that time believed in a racial theory of mankind that was based on colour alone. They saw themselves as belonging to a superior 'white' or Caucasian race.

OPTION TOPIC 19
THE IMPACT OF COLONISATION ON MIGRATION AND RACIAL PATTERNS

COLONISATION OF THE AMERICAS

◀ South America, showing Dutch, French, Spanish and Portuguese regions

- The colonial powers were European countries who wanted to increase their wealth by sourcing raw materials, such as spices or products like cotton, sugar or tobacco, for their industries at home. So most of the New World was colonised by Caucasians (white Europeans).
- The Spanish, Portuguese, British and French were the main colonisers in the Americas.
- Black slaves from West Africa were transported to work in the plantations of the Americas and the Caribbean islands. This led to people of black African origin forming part of the American population.
- The majority of African-Americans live on the south-east and south coasts, the regions where slavery was mostly practised.
- The Spanish colonised Mexico and the regions to the south, including all areas of Latin America except Brazil, which was settled by the Portuguese under the Treaty of Tordesillas.

- The Spanish explorers Cortes, who conquered the Aztecs, and Pizarro, who conquered the Incas, brought their Spanish culture, their language and architecture, a part of the colonising process.
- The Spanish colonised the west-coast region of the USA and many of its towns and cities bear Spanish names, such as Las Vegas, San Francisco, San Diego and Santa Barbara.
- A large percentage of people who live in the South West of the USA are of Spanish extraction. Up to 98 per cent of large parts of western Texas is Hispanic.
- Black slaves were brought to the south-east and south to work on the plantations.

■ South-East Asia

- Throughout past centuries people emigrated from China in search of a better living standard that they had at home.
- The land south of China is a region of mountain ridges divided by deep valleys that provided easy, ready-made routes for migrants. So a large percentage of these countries have Chinese ancestry.

■ Australia

- Australia was a British colony, so the majority of the population is Caucasian.

■ Africa

- South Africa was settled by the Dutch and British at various times. While the majority population is black, a small population is of Caucasian extraction.
- Other parts of Africa were colonised by other European countries, such as Belgium and Spain.

OPTION TOPIC 20
LANGUAGE AS A CULTURAL INDICATOR

- 'Cultural regions' is the general term for areas where some portion of the population shares some degree of cultural identity.
- Language is clearly evident in such regions through place names.
- The choice of language on signs is another visible symbol of culture in the landscape.

- **Case Study: Euskara – the Language of the Basques**
- The Basque language forms a crucial part of their unique identity.
- The language of the Basques is called **Euskara.** It is spoken by about 520,000 people.
- It is one of the oldest living languages and is not known to be related to any other language.
- It was spoken in the Basque region in Neolithic or Stone Age times.
- The first written texts in Basque date from the tenth century.
- Basque was a forbidden language after the Spanish Civil War in the 1930s, during the reign of the dictator General Franco.
- Basque schools called 'iskastolas' started in the 1930s in defiance of this policy.
- Because there were many dialects of Basque, steps were taken in 1964 to create a unified Basque language.

THE BASQUES – A CULTURAL GROUP WITHOUT NATIONALITY

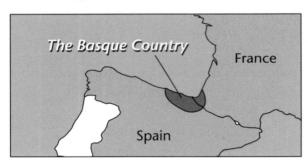

◀ Part of the Basque region is in Spain. The other, smaller part is in France.

- The Basque country is a region at the western end of the Pyrenees, the mountains that divide France from Spain.
- It is made up of seven districts, four of which are in Spain and form the largest section, while the other three are in France.
- Three of these historic Basque territories – Araba, Bizkaia and Gipuzkoa in the north of Spain – are grouped together to form a political unit, known as **Euskadi**, or the Autonomous Community of the Basque Country.
- **Euskadi** has a population of 2.1 million people. They have their own president and parliament but are represented internationally by Spain.

Spain's other Basque district, Navarra, is its own region, separate from Euskadi and less troubled politically.

■ **Who are the Basques and why are they Different?**

The Basques were living in the Pyrenees over 4,000 years ago, long before the Celtic tribes of central Europe moved west to Britain and Ireland.

- They speak Basque, which they call Euskara, a language unrelated to any other human tongue.
- Basques also speak Spanish or French. Some speak both.
- Basque cuisine is based on seafood, especially cod and hake.
- Basque dishes are very popular throughout Spain and most major cities have Basque restaurants.
- There is an event that has made Pamplona, a city in Navarra, famous. Every year, six bulls are allowed to run freely through Pamplona's streets before being killed later that day by matadors in a bullring.
- Many people run ahead of the bulls and sometimes get hurt or killed.

■ **Basque Conflict with the Spanish Government**

- A small number of violent extremists are represented by ETA.
- **ETA** (in Basque it means Basque Homeland and Freedom) **is an armed nationalist group**.
- They believe that complete independence from Spain and France can be achieved only by military means – similarly to what the IRA believed about the reunification of Ireland.
- Initially ETA was founded in 1958 because the Basque people were oppressed during the reign of the fascist dictator Franco in Spain.
- In the beginning they were a non-violent group, but their every move for independence was put down by force. This made them opt for armed resistance.
- ETA is not represented in power sharing in government. So ETA has returned to violence, the very method it discarded as hopeless in 1998.
- A key concept of ETA is the 're-nationalisation of the Basque country' – to restore the Basque region to its full cultural personality.
- Basque nationalism has risen in France, an area that was originally stable. Policy changes on social conditions and welfare and the loss of influence by trade unions has led to discontent.

OPTION TOPIC 21
THE MAJOR LANGUAGE FAMILIES

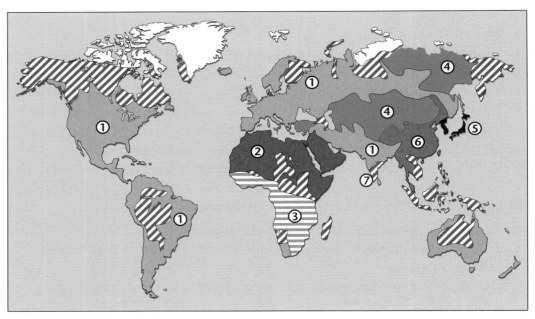

🔺 The major languages of the world. Use the map above to name the language families 1–7.

■ **The Indo-European Family**

- About half the world's population speaks languages from this family.
- It began in the region now called Turkey; its speakers migrated to various regions and the language changed along the way.
- The colonisation of the New World by the European colonial powers helped in the spread of this language family.

■ **The Sino-Tibetan Family**

- This is the second most widely spoken language, with over 1 billion speakers.
- It includes Chinese, Thai, Burmese and Tibetan.

■ Arabic–Semitic Family

- This language family includes Arabic and Hebrew.
- It is mostly spoken in north and north-east Africa, the Middle East and the Arabian peninsula.
- The spread of the Muslim faith across Africa brought Arabic to this region.

■ **The Ural and Altaic Family**

- This family includes Finnish, Hungarian and Turkish, as well as the languages spoken in the Asian part of Russia.

■ **Niger–Congo family**

- This is also called Bantu.
- This region stretches from the Sudan in the north to South Africa in the south.
- It includes Swahili, which developed as a pidgin language from contact with Arabic traders along the east African coastline.

■ **Japanese and Korean family**

- This family is limited to Japan, North Korea and South Korea.

■ **Dravidian family**

- This is spoken in southern India and Sri Lanka.
- It includes Tamil.

OPTION TOPIC 22
GAELTACHT REGIONS

- In 1925 Gaeltacht regions were divided into two categories:
  - **Fior Gaeltacht** regions, where 80 per cent or more of their population speaks Irish
  - **Breac Gaeltacht** regions, where 25 to 79 per cent of their population speaks Irish.

At this time Gaeltacht regions covered substantial areas of the West. Today, however, Gaeltacht regions have reduced in size and number and are confined to scattered regions along the west and south coasts. They have a total population of about 86,000 people.

INITIATIVES FOR THE SURVIVAL OF THE IRISH LANGUAGE

- Festivals that promote the language through art exhibitions and music
- An audio-visual industry that promotes Irish culture within the Gaeltacht and throughout Ireland. These influences include:
  (a) Radio na Gaeltachta
  (b) The TV station TG4

- Local radio stations such as Leirithe Lunasa Teo in Corca Dhuibhne and Nemeton Teo in An Rinn
- Irish-language schools, na Gaelscoileanna
- Summer colleges in the Gaeltacht areas.

■ Supports for Survival

- Article 22 of the European Charter of Fundamental Rights states that the Union respects cultural, religious and linguistic diversity.
- The European Bureau of Lesser-used Languages (EBLUL) works on behalf of those in the EU who speak minority languages. It creates and supports policies that support these languages.
- The EU gives financial aid to EBLUL.
- International conferences are held to identify ways to improve the situations of these minority languages. These are organised by the Foundation for Endangered languages (FEL).

OPTION TOPIC 23
THE DISTRIBUTION OF THE MAIN RELIGIONS OF THE WORLD

■ How Religion Acts as a Cultural Indicator

- Religion creates landscapes by the construction of religious buildings.
- There are churches in Christian regions.
- Mosques with their minaret towers are found in Muslim areas.
- There is an absence of bars and pig farms in Muslim areas.

Personal indicators

- Turbans are worn by Sikh men, who also have long beards.
- Chadors are worn by Muslim women.
- There are differing attitudes towards women, to birth control, to materialism.

THE WORLD'S MAJOR RELIGIONS

■ Judaism

- There are only about 14 million Jewish people.
- It is a religion with its origin traced back to Abraham, who migrated from Mesopotamia, now called Iraq, to Palestine with his followers.

- Issac, who was Abraham's true heir, became the ancestor of the Israelite people.
- Jews think of Palestine as their spiritual homeland.
- **Israel was founded in 1947** to provide a secure homeland for the Jewish people.
- Since then Jews have travelled from all over the world to Israel.
- Any immigrant Jew to Israel and his wife are entitled to Israeli citizenship.

■ Christianity

- Christianity is the largest religious group, with over 33 per cent of the world's population as believers.
- Christianity has its origins in Judaism.
- Christians believe that Jesus Christ was the messiah prophesied in the Old Testament.
- Christianity spread rapidly through the work of St Paul and other missionaries.
- The Romans persecuted the Christians for many years until the Emperor Constantine granted them freedom of religion.
- Because of the Edict of Milan in AD 313, Christianity became the official religion of the Roman Empire.

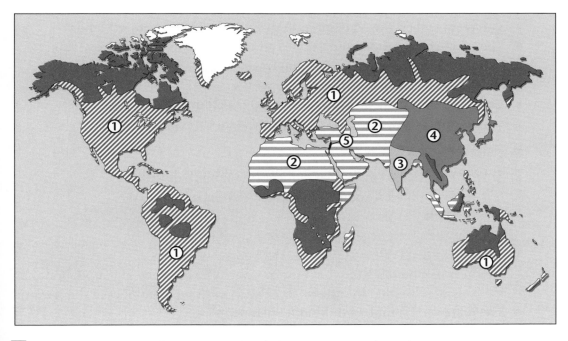

▲ Identify the main religions in each of the regions numbered 1 to 5.

- The rulers of Spain and Portugal were Christian and were the main cause of the Christian faith's spread to Latin America.
- The British and French, also colonial powers, spread the Christian faith to the USA, Canada and Australia.
- The Jesuits carried Christianity to China.
- Christianity had spread throughout Russia before the Russian Revolution.

■ Islam

- Islam is the name given to the religion preached by the Prophet Mohammed in the seventh century.
- Mohammed was an Arab who was born in Mecca about AD 570.
- Mohammed preached that there was only one god, Allah.
- Those who believe in this faith are called Muslims.
- Islam has its origins in Judaism and Christianity.
- Mohammed believed that Christ was a prophet of Islam.
- He was forced to flee to Medina in AD 622 and this date marks the beginning of the Muslim calendar.
- Mecca and Medina are the most important sacred cities of Islam.
- After the death of Mohammed, the new Caliph, or ruler, and his successor waged holy wars known as Jihads.
- These wars captured all the lands of North Africa and parts of Europe. They ended at the Battle of Tours.
- The wars spread the Muslim faith throughout their captured lands.
- The Arab conquerors taught the Arabic language to all their new subjects and it replaced all their native languages.
- Islam has two branches, Sunni and Shiite.
- Sunnis make up about 84 per cent of all Muslims.
- Shiites are confined mainly to Iraq and Iran.

■ Hinduism

- Mostly confined to the Indian subcontinent and South-East Asia.
- It is not promoted through missionary activities.
- Hinduism traces its origin to tribes of Indo-European migrants who brought their language, Indo-European, with them.
- The sacred texts of Hinduism are the Vedas.
- Hindus are monotheists, which means they believe in one high god, Brahman.
- Hinduism is divided into three sects, each with its own view of the nature and name of the high god: Visnu, Shiva or Shakti.
- Educated Hindus believe that the three gods are merely differing ways of looking at the same high god.

- The Ganges is the sacred river of the Hindus.
- They believe that its waters are immaculate and nothing can pollute it.

■ Buddhism

- Buddhism is based on the teachings of Siddhartha Gautama.
- He was born in about 563 BC.
- Buddhism began near the borders of India and Nepal.
- Lumbini, now in Nepal, and Bodh Gaya and Kusinagara in India are the sacred places of Buddhism.
- Missionaries spread this faith to China, Tibet, Vietnam, Korea and Japan

OPTION TOPIC 24
THE RELATIONSHIP BETWEEN CHURCH AND STATE

THE IRISH CONSTITUTION

1. The Irish Constitution was introduced in 1937.
2. The Constitution:
   - recognised the special position of the Catholic Church
   - guaranteed freedom of conscience and the free profession and practice of all religions.
3. The power of the Catholic Church was used to influence government decisions such as the withdrawal of the 'Mother and Child' scheme.

■ Northern Ireland

- 62 per cent of Church of Ireland members and almost all Presbyterians lived in the same nine Ulster counties.
- Political control was predominantly in the hands of Presbyterians.
- The Plantation of Ulster gave rise to the large Presbyterian population and the spread of Calvinism.
- The Orange Order represented the wealthy landowners, industrialists and the Presbyterian community.
- Social segregation of Catholic communities from Protestant communities led to ghettos in city regions.
- Gerrymandering was the arrangement of voting whereby only those who held property were entitled to vote for local councils, and those who had many properties had many votes.

- There was widespread corruption and discrimination against Catholics.
- Civil Rights marches were organised by the Catholic communities to get equal treatment with Protestants.

OPTION TOPIC 25

RELIGIOUS CONFLICT

CASE STUDY 1: RELIGIOUS CONFLICT BETWEEN INDIA AND PAKISTAN OVER THE KASHMIR REGION

See pages 127–8.

CASE STUDY 2: RELIGIONS CONFLICT IN NORTHERN IRELAND

- Between 1966 and 2000, over 3,600 people have been killed and 36,000 wounded as the conflict spread into mainland Britain and the Republic of Ireland.
- This period is known as 'The Troubles'.
- Since 1997, a ceasefire has held among the main paramilitary groups, such as the IRA, the UDA and the UFF.
- The Good Friday Agreement was signed on 10 April 1998.
- The Former US Senator, George Mitchell, was chairman of the all-party talks that led to the agreement.
- Tension has existed between the two faiths since the reign of Henry VIII (1509–49), when the Protestant faith was introduced to Ireland.
- The Treaty of 1921 recognised this religious division by dividing the country into a predominantly Protestant Northern Ireland and a predominantly Catholic Republic of Ireland.
- The Good Friday Agreement created a 108-member assembly and a fourteen-member executive body in which both Catholic and Protestant members sit together in government.
- Paramilitary groups have for the moment laid aside their weapons.

OR

THE PALESTINIAN AND JEWISH CONFLICT IN THE MIDDLE EAST

- The Jews and Arabs both trace their ancestry back to Abraham.
- Jews who migrated throughout the world think of Palestine as their spiritual birthplace, because their religion developed there.
- Both groups claim Israel.

- The Jews claim it because of God's promise, upon which they developed their religion, and because they have lived there for thousands of years.
- The Arabs claim Israel because Palestine became chiefly an Arab land after they conquered it in the sixth century AD, as a consequence of conquest,colonisation and the spread of Islam (see Islam, page 258), and because Arabs have lived there as the majority population ever since.

■ The Zionist Movement

From the mid-1800s the Turks controlled this region, and Jewish settlers returned to Palestine in order to live and die in the Holy Land. Some Jews formed a movement called Zionism, which sought to make Palestine an independent Jewish nation.

- From the 1800s until World War One, most of the world was controlled by empires. These included the British Empire, the French Empire and the Ottoman (Turks/Turkey) Empire.
- During World War Two the Turks sided with Germany against the Allies.
- Because of Arab support during World War One in fighting the Turks, the Allies promised to back Arab demands for independence in the Middle East.
- Great Britain and its European allies planned to divide the Ottoman Empire up into nations after the war; Britain was to take control of Palestine to create a homeland for Jews and for the Arabs who already lived in Palestine. Palestine was to be independent and governed by 'self-governing institutions' that would cater for the rights of all minority groups.
- However, the terms of the mandate (agreement) were not clear and Jews and Arabs interpreted it differently.
- The Arabs believed that Palestine was promised to them for their support against the Turks in World War One, while the Jews believed that Palestine was to be a homeland for the Jewish people.
- Then during the 1930s the Nazi persecution of Jews by Germany brought large numbers of Jewish refugees to Palestine.
- The Arabs revolted against British control because the British allowed this immigration.
- In 1939 Britain decided to end Jewish immigration within five years. The Jews opposed this.
- However, both sides put aside their differences during World War Two.
- The Nazis killed about 6 million European Jews.
- Britain continued to limit Jewish immigration into Palestine and protesting Jews organised large-scale demonstrations.

- Finally, with no solution in sight, Britain asked the United Nations to solve this problem of religious conflict.
- **The United Nations Special Commission on Palestine recommended that Palestine, which lay to the west of the river Jordan, be divided into an Arab state and a Jewish State.**
- The Jews accepted this decision but the Arabs did not. The Jewish people got slightly more than half of Palestine, including of course land that had long been occupied by Arabs.
- The previous year, in 1946, the British granted independence to the territory lying east of the river Jordan and 'Transjordan' (now the state of Jordan) came into existence.
- **On 14 May 1948 the Jews proclaimed the independent state of Israel** and the British withdrew from Palestine.
- **The next day neighbouring Arab nations that were part of the former Ottoman Empire, such as Syria and Jordan, together with Lebanon and Egypt that had not been part of the Ottoman Empire, invaded Israel in an attempt to help the Palestinian Arabs destroy the new Jewish state. These Arab nations rejected Israel's right to exist.**
- When the fighting ended Israel held territory beyond the boundaries provided by the UN plan and controlled 80 per cent of Palestine.
- **During the conflict, Transjordan, the new Arab state to the east of the river, invaded the Arab region of Palestine called the West Bank, and claimed it for itself. Its leader renamed it Jordan.**
- More than 600,000 Arabs who lived within Israel's new borders fled from the Jewish state and became refugees in neighbouring Arab countries.
- This early conflict was only the first in a series of wars between Israel and its Arab neighbours.
- **In 1967, a week-long war resulted in a major Israeli victory: Israel took the Golan Heights from Syria, the West Bank from Jordan and the Sinai Peninsula up to the Suez Canal from Egypt. In 1973, another brief war led to Israel's withdrawal from the Suez Canal to truce lines in the Sinai Peninsula.**
- In 1978, Israel made peace with Egypt and Jordan and it later returned all of the conquered Sinai Peninsula to Egypt. In the 1990s some progress was made on key issues involving Israel and its Arab neighbours, and between Israel and the leadership of the Palestinians who have been under Israeli control in the Gaza strip and in the West Bank.
- Agreements creating peaceful relationships with Egypt and Jordan were major steps on this difficult road, and the establishment of the Palestinian

Authority to govern Gaza as well as several West bank cities was another achievement.
- Many obstacles still stand in the way, however, including the following:
  - new territories with housing estates ('settlements') built by Israelis on Arab lands
  - the concept of Jerusalem as an open city for Christians, Muslims and Jews
  - concrete walls built to divide Arab regions from Israeli lands.

OPTION TOPIC 26
NATIONALITY

- Nation state refers to a country that occupies a specific area of land, and this area is occupied by a national group who share a common culture.

The concept of nation state combines three elements:
Nation (or ethnicity)
State: the type of government or regime in power
Territory: the area defined on the ground that is controlled by the state.

NATION

A nation is a group of people who feel bound together through personal ties and who possess a unity and solidarity that has grown by:
- following a common way of life
- sharing common experiences
- possessing common cultural traits
- inheriting a common tradition.

- Nationalism is the cause through which such groups claim their right to be a sovereign power within a particular area of land.
- Nations rarely consist of just one ethnic group.
- The factors that create and maintain national feeling include:
  - ethnic group or race
  - language
  - religion
  - a common enemy.
- A state boundary sometimes coincides with a physical natural barrier, such as a mountain range like the Pyrenees, or a river like the Rhine between France and Germany.

- A frontier is a zone or area which separates one ethnic group or nation from another, for example the 'Border' between Northern Ireland and the 'Irish Republic'.

■ The Influence of Shape, Size and Location of a Nation State

Shape
- A circular shape is best. It has a minimum boundary to protect, and its capital can be equidistant from all boundary locations.
- Long countries will have core regions and peripheral regions, such as in Italy. The north-west is a core region; the Mezzogiorno is a peripheral region.
- Some nations may be formed of one island, such as Ireland, or many islands, such as Indonesia.

Size
- A large state may create a greater sense of security and have a greater variety of resources than a small state. Small states are more vulnerable in times of war and are easily overrun by enemy forces.

Location
- A country's absolute location is its position on the globe.
- Its relative location may change from time to time depending on:
  – types of transport
  – location of its main trading blocs.

■ Everyday Expressions of Irish Culture and Identity

Drama
- Many plays have been written that are based on Irish culture and folklore. 'The Playboy of the Western World' and 'The Field' are just two of many such Irish dramas that have been performed in the Abbey Theatre in Dublin and at other venues throughout the country.

Sport
- The Gaelic Athletic Association has promoted Gaelic games and has the greatest membership for all sports in Ireland.
- Every parish has its own GAA club.
- Their most important events are the two All-Ireland Finals that are held in September each year.

Music and dance
- Traditional Irish music and dance are popular expressions of Irish culture.
- Feis Ceoil competitions are held regularly.
- The designs on female costumes for Irish dancing are based on Celtic patterns.

Festivals

- The St Patrick's Day parade is held each year on 17 March in all our major towns. The largest parade is held in New York, where a large population of Irish people and people of Irish descent live.
- The Twelfth of July festival celebrates the identity of the Unionist community in Northern Ireland.
- Many small towns have special events that celebrate other interests of Irish people, such as the Wexford Opera and Arts festival.

Summer schools

- Summer schools promote learning and interaction among students, poets, writers, politicians and university lecturers in an informal setting.

■ Festivals and Sports in Europe

- The 'Running of the Bulls' is an annual event in Pamplona, in the Basque region of Spain.
- The Munich Beer and Music festival is the largest festival in the world, which over 6 million visitors visit each year.
- Ice-skating and skiing are popular sports in snow-covered upland regions such as the French, Austrian and Swiss Alps, and include cross-country skiing, especially in Scandinavia.
- The Tour de France is the most famous of all French sporting events.

OPTION TOPIC 27
ISSUES RELATING TO PHYSICAL AND POLITICAL BOUNDARIES

WATER SUPPLIES

- The probability of conflicts over water supplies is great, especially in times of water shortages.
- The rivers that flow through some nations do not rise in those nations. For example, the river Indus in Pakistan rises in India. India takes some water from the river before it flows into Pakistan. If too much is taken then less flows through Pakistan, much of which is desert and relies heavily on this source for its water supply.

OFFSHORE BOUNDARIES

- The water, the seabed and their resources – such as oil and gas – within **200 nautical miles** of a country's seashore belong to that country.
- This area is called the country's Exclusive Economic Zone.
- Territorial seas extend up to **12 nautical miles** offshore. States must allow the innocent passage of foreign ships through these waters.
- Where states adjoin each other, lines halfway between the nearest shorelines of each state must be decided.
- The political boundaries of a state define its ability to enforce its laws.

POLITICAL BOUNDARIES

■ Political Boundaries and Ethnic Divisions

Political boundaries often divide an individual ethnic group into two or more divisions.

- The Basque region is divided into two parts. One part, the larger, lies in Spain. The other, smaller part lies in France. They are separated by the Pyrenees mountain range that forms the boundary between France and Spain.
- The border area separates the nationalist people in Northern Ireland from their southern cultural part that lies in the Republic of Ireland.

OPTION TOPIC 28

CULTURAL GROUPS WITHIN NATION STATES

NATIONALISTS IN NORTHERN IRELAND

- People who live in Northern Ireland and wish to be part of a united Ireland are called nationalists.
- Since partition in 1921, nationalists have lived under British rule and have suffered discrimination.
- Nationalists are Irish in their traditions and customs and most are Catholic.
- Nationalist feelings developed as a consequence of the British occupation of the island of Ireland and the oppressive treatment of the Catholic minority for over seventy years.
- The separation by fear of Protestant and Catholic communities has led to the development of ghettos in Derry and Belfast.
- Demands for civil rights and equal treatment with Protestant citizens led to civil-rights marches in the 1960s and 1970s. Thirteen civilians were shot dead in a civil-rights march on 'Bloody Sunday' in January 1972 by the British forces.

- A minority of extreme nationalists support the IRA, an illegal, armed, paramilitary organisation.

  ### ■ The IRA and the British Government
  - As a consequence of Bloody Sunday the IRA focused and intensified its campaign of violence on British cities and army barracks.
  - Internment without trial of IRA sympathisers intensified the campaign further.
  - Ten IRA prisoners died in prison to highlight their claim for political rather than criminal status.
  - Efforts by SDLP leader John Hume and Sinn Féin leader Gerry Adams led to a ceasefire that continues to this day.

TURKS IN GERMANY

- There are 1.8 million Turks in Germany, with 139,000 Turks in the city of Berlin alone.
- They were recruited from 1961 onwards to do low-paid jobs in Germany that the educated Germans would not do.
- The migrants dreamed of earning enough money to start up a business of their own in Turkey.
- The population of Turks in Germany remains high because many chose to remain in Germany and the birth rate among Turks is high.
- Many Turks of the second and third generations, born or raised in Germany, have little knowledge of Turkey.
- Turkey has a similar status to German-born Turks as Ireland has for Irish-Americans.
- The German-born Turkish group have little desire to go to live in Turkey, a country they are unfamiliar with.
- There is a high number of school-going German youths of Turkish origin. This increases educational cost for the state, as their home language is Turkish and their German is not fluent.

OR

BASQUES IN FRANCE AND SPAIN
See pages 252–3.

OPTION TOPIC 29
CULTURAL GROUPS WITHOUT NATIONALITY

■ **Nationalists in Northern Ireland**
See pages 266–7.

■ **Basques in Spain**
See pages 252–3.

OR

SIKHS IN INDIA

- The Sikhs form a cultural group that belongs to a religion founded by Guru Nanak about 500 years ago.
- Guru Nanak tried to unite Muslims and Hindus of all castes into a single faith.
- Most followers of the Punjab region are Sikhs and followers of Guru Nanak.
- Their holiest shrine is the Golden Temple in Amritsar.
- Many Sikhs earned positions of trust in the defence forces during British colonial rule. This gave them a sense of middle-class status when independence came in 1947.
- All Sikhs resemble each other by wearing five symbols, called the 'K' symbols. They are: Kesh, Kangha, Karra, Kachha, Kirpaan.
- Sikh male adults wear a dastar, or turban, and have long beards.
- Sikh women cover their heads with a long scarf, called a chunni.
- Sikhs who follow all these traditions are called Khasla. People who follow only some of these conditions are called Sahajdharis.
- This symbol of distinct identity, defined by dress code, is called 'bana'.
- Many Sikhs seek full independence from India.
- To satisfy some of their demands the Indian government made the Punjab a separate, Panjabi-speaking state where the Sikhs are the majority rulers.
- Many Sikhs want full independence, in a state they would call Khalistan.

OR

THE KURDS

- The Kurds are mainly a Sunni Moslem people with their own language and customs.
- Most Kurds live in south-west Asia in the mountainous region that forms part of Turkey, Iraq, Iran, Armenia and Syria. This region is known as Kurdistan.

- Before World War One traditional Kurdish life was nomadic and they were herders of sheep and goats.
- The break-up of the Ottoman Empire after World War One created a number of new states, but not a separate Kurdistan. Kurds, no longer free to roam, were forced to abandon their seasonal migrations and traditional ways.
- The 1920 Treaty of Sèvres, which created the modern states of Iraq, Syria and Kuwait, was to have included a separate Kurdistan; but that was never implemented.
- Turkey persecuted the Kurds and tried to deprive them of their identity, their language and their traditional dress.
- Turkey still does not recognise them as a cultural group.
- Kurds in other countries have also been suppressed, such as in Iraq under the power of Saddam Hussein.
- The Kurds rebelled in Iraq after the Gulf war, but their efforts were crushed. Chemical weapons were used by Saddam Hussein against the Kurds.
- The United States has tried to create a safe haven for the Kurds by imposing a 'no-fly zone' north of the 36th parallel.
- The Kurds are divided amongst themselves. Some might accept a federal status within other countries; many, however, will not. Numerous violent protests have taken place between Kurdish groups over their quest for independence.

THE PALESTINIANS

- The Palestinians are an Arab people who have lived in south-west Asia for thousands of years.
- Their religion, customs and traditions are different from those of the Israeli people.
- They were given some lands in various separated units in 1947, when the territories of the Ottoman Empire were divided after World War Two.
- This division led to the creation of two states in Palestine: (a) a new Palestine State and (b) Israel.
- The Palestinians lost their national territory, because they and the neighbouring Arab states did not recognise Israel's right to exist. They attacked the new state of Israel in 1948 and as a consequence set in motion a sequence of events that have left them without any independent state of their own.
- Since then Arabs who had called Palestine their homeland for centuries have lived as refugees in neighbouring countries.
- Israel is now more than 50 years old and most Palestinians were born after the partition of Palestine in 1948.
- The Palestinians call themselves a nation without a state (much as the Jews did before Israel was founded), although they and their descendants make up the majority of Jordan's population today.

- They demand that their grievances be heard and that a Palestinian state be created.
- The first steps towards such a state have been taken as part of the Arab–Israeli peace process, and these negotiations continue. Current estimates of Palestinians in the Arab world total about 8.5 million people.

OPTION TOPIC 30
IDENTITY

Identity involves a variety of cultural factors, such as nationality, language, race and religion.

CASE STUDY: GERMANY

Its natural boundaries include:
- the Alps in the south
- the River Rhine and the Eifel Uplands, dividing France from Germany
- the Bohemian Forest and Erzgebirge uplands, separating Germany from the Czech Republic
- The River Oder, separating Germany from Poland.

■ Religious Conflict within Germany

- The Reformation began in Germany.
- Martin Luther published his theses in 1517.
- The German Empire began in 1871.
- One-third of its people were Catholic and two-thirds were Protestant.
- German Catholics organised themselves into a new party called 'The Centre Party', which opposed the Empire.
- The Pope proclaimed infallibility to be a doctrine of faith and some Catholics did not believe this could be true.
- The Prime Minister, Bismarck, used this division among Catholics to crush all opposition to his empire.
- The May Laws were used to reduce the power of the Catholic Church in German affairs.
- The powers of the clergy were restricted and priests were forbidden to raise political topics in their sermons.
- Civil marriages were made compulsory; many priests were put in prison and many others had to leave the country.
- A considerable number of these laws were withdrawn after 1879, when a new pope was elected.

■ Old and New Boundaries

- Before World War One Germany was a larger state than it is today. It included all Alsace-Lorraine in France and all the north-western and western regions of Poland.
- After the war the German Empire had fallen and the country lost Alsace-Lorraine and all its lands in Poland. New boundary lines were drawn to reflect these changes.
- Adolf Hitler was violent and ambitious. He wanted to unite all German speakers into a single nation state in a greater Germany.
- He regarded Germans as a superior 'master race'. He wanted to be rid of all others, whom he felt were inferior.
- He wanted to create more living space for Germans. He believed that the Treaty of Versailles had robbed Germany of much of her lands.
- When the German army invaded Poland in 1939 World War Two began.
- At the end of World War Two, new political boundaries were drawn in Germany.
- Germany was divided by the Allied Powers.
- Three zones occupied by Western powers were united to form West Germany.
- The USSR created a new state in East Germany, called the German Democratic Republic, from the zone that it controlled.
- The city of Berlin was also divided into Western and Russian zones by the Berlin Wall.
- In 1990 East and West Germany were reunited.

Consequences of changing political boundaries

- Many German-speaking areas now lie outside the nation state of Germany.
- People whose ancestors were born within the German state in the past may find their relations are citizens of a new state, even though they may never have moved from their ancestral homeland.
- World Wars One and Two caused the deaths of millions of people.
- Many cities occupied by Germany and within Germany itself were badly damaged during these wars.
- Germany was divided after World War Two into two separate states: the German Democratic Republic and the Federal Republic of Germany. It was not to be reunited again until 1990.

■ Ethnicity and Race

- Hitler encouraged division of people according to race.
- He believed in an Aryan Race of Germans.
- He persecuted Jews, gypsies and homosexuals.

Ethnic groups in Germany

- Ethnic Germans are descendants of Germans who lived in lands in Eastern Europe and Russia that had at one time been part of Germany.
- They held rights to German citizenship according to its constitution and once within Germany they automatically became citizens, even though they may not be able to speak German now.
- The number of ethnic Germans swelled from 40,000 annually in the 1980s to about 200,000 annually in the 1990s.
- This created a strain on German resources for health benefit, education and housing.
- Under new legislation new immigrant ethnic Germans must live within certain areas. If they do not they lose benefits.
- To many Germans, ethnic Germans do not seem to be German.

Ossies and Wessies

- West Germans are colloquially called Wessies and East Germans are called Ossies.
- After reunification many Germans from the former East Germany migrated to West Germany.
- Ossies feel that Wessies regard them as inferior, and many feel they are treated as second-class citizens.
- Wessies resent the fact that their taxes were increased to support the unemployed Ossies.
- Wessies also feel that Ossies are lazy and ungrateful.

■ Music in Germany

- Johann Sebastian Bach, Georg Handel, Felix Mendelssohn and Richard Wagner are some of the famous German composers of classical music, although Handel later came to live in England.
- The Munich Oktoberfest celebrates German traditional music.

■ Sport in Germany

- One in three Germans, or 26.6 million people, are members of sports clubs.
- Fifty-eight per cent of German people who regularly participate in sport are not members of sports clubs.
- Thirty-seven per cent of Germans take part in sporting activities at least once a week.

OPTION 3: GEOECOLOGY

OPTION TOPIC 31
SOIL COMPOSITION AND CHARACTERISTICS

SOIL COMPOSITION

The scientific study of soils is known as **pedology**.

Soil is the surface layer of loose material that covers much of the Earth's land surface. It contains both organic and inorganic matter.

It is part of the natural environment that links the relationship between bedrock, climate and vegetation.

■ Composition of Soil

All soils contain mineral particles, humus, water, air, and living organisms such as bacteria.

Mineral matter

- A soil gets its mineral content from its parent material. The parent material can be bedrock which is broken down by physical and chemical action, glacial deposits, river deposits or wind-blown deposits.
- Mineral content refers to minerals such as calcium, phosphorus, potassium, potash and other compounds.
- These are the foods that plants need in order to grow.
- The parent material determines the soil colour, depth, texture and pH value.

Organic matter

- Organic matter is also referred to as humus.
- Humus forms from decayed plants, and to a lesser degree from animal life through the action of bacteria and other micro-organisms.
- Humus also improves the texture of a soil.
- Humus binds soil particles together, which increases the soil's ability to hold moisture.
- Plant roots help to bind soil particles together.

Climate

- The distribution of the various soils coincides with the distribution of the world's climate.
- So climate influences the type of vegetation that grows in a region.
- The greater the amount of vegetation that grows in a region, the greater will be the humus content of the soil.
- Climate influences the rate at which weathering of the soil and decay of its plant matter occur.
- Heavy rainfall causes leaching of the soil's minerals.
- Weathering is the first state of soil formation.

Slope and water movement

- Steep slopes encourage the removal of fine particles by rainwater.
- Gentle and flat slopes encourage the accumulation of fine particles.
- When rainfall is greater than evaporation water moves from the surface layer to lower layers and carries down nutrients with it. This process is called leaching.
- The downward movement of water may create a hard pan that can prevent drainage and lead to waterlogging.

Air

- Air is vital in the soil for oxidation, which converts parts of the organic matter into oxygen.
- Air is vital also for the bacteria present in the soil, which require oxygen and are said to be aerobic.

■ Soil Characteristics

Soil texture

- Texture refers to the proportions of sand, silt and clay particles that make up the soil.
- Texture determines the soil's ability to:
  1. retain and transmit moisture
  2. retain nutrients
  3. allow roots to penetrate it.
- Sandy soils consist of 70 per cent or more sand particles by weight and have few nutrients.
- Clay soils consist of 50 per cent clay particles by weight. Clay soils are rich in nutrients, but are likely to become waterlogged.
- Silty soils are intermediate between sandy and clay soils.
- Loam soils are ideal for agriculture. They contain a mixture of particles of many different sizes.

- Loamy soils are well aerated, with some moisture and plant food.
- Sandy soils feel gritty to rub.
- Silty soils have a smoother, soap-like feel.
- Clay soils feel sticky or plastic when wet.

Colour

- Humus-rich soils are dark brown or black in colour.
- Brown soils of temperate forest lands, such as Europe, get their colour from decayed leaves and plant particles.
- Organic matter, such as leaves, in rainforest soils decays rapidly owing to the high humidity of the forest floor.
- Dark soils absorb more sunlight and so are warmer than light-coloured soils.
- Warm soils aid germination of seeds and have a long growing season.
- Soil texture can be determined by examining a soil sample. This can be done by mixing a soil sample with water in a jar.

Soil structure

- Soil structure refers to the shape of the soil grains or particles.
- In undisturbed soils the clustering of particles forms different shapes known as peds.
- The shape and alignment of the peds, together with their size, determine the size and number of pore spaces through which air, water and organisms can pass.
- Soils with a crumb structure give the highest agricultural yield.
- A crumb structure is best because it provides the best balance between air, water and nutrients.

Water content

- Water moves downwards by percolation and upwards by capillary attraction.
- Water in the soil becomes a weak solution of many mineral compounds.
- The chemical processes that take place in a soil do so mainly in solution.
- Water content varies between soils, from almost nil in arid climates to waterlogging in wet clay soils.
- Constant percolation of water downwards in mid to high latitudes, owing to high rainfall, causes leaching.
- Leaching is a process whereby minerals are drawn downwards from the upper horizon at the surface, to lower horizons.
- Leaching creates podzol soils.

Organic content

- Organic matter includes humus, and is formed mostly from decaying plants and animals.
- Humus gives the soil a dark colour.
- The highest amounts of organic matter are found in the chernozems or black earths of the North American prairies, Russian steppes and Argentinian pampas.
- Fallen leaves and decaying grasses and roots are the main sources of organic matter.
- Soil organisms, such as bacteria and fungi, break down the organic matter.
- Where soil organisms are present and active they will mix the plant litter into the A horizon, where it decomposes into humus.

pH value

- A soil's pH is a measure which grades it as either acid or alkaline.
- A soil with a low pH value is said to be acidic. This happens when minerals such as calcium, magnesium and potassium are leached by heavy rain.
- A very low pH, or acid soil, slows down decomposition and may even prevent decay, as in the creation of peat.
- A high pH means an alkaline soil. This indicates it has a high calcium (lime), magnesium and potassium content.

OPTION TOPIC 32
FACTORS THAT AFFECT SOIL FORMATION

■ Parent Material

- If the soil is composed of unconsolidated deposits such as boulder clay, soil formation will occur more rapidly than if it were bedrock.
- Soils inherit their characteristics from parent material.
- Soil from limestone areas will have calcium and other minerals and will become alkaline.
- Soils from sandstone regions will be free-draining and sandy and will heat up quickly in spring.
- Soils from clay regions will have poor drainage and may cause waterlogging.

■ Climate

Effects of temperature and precipitation include:

- Constant heat and moisture encourage constant growth.
- Soils in hot, wet regions tend to have lots of vegetative cover, such as rainforest or monsoon forest.
- Decomposition occurs rapidly.
- Chemical weathering occurs rapidly, creating deep-red soils.
- Heavy rainfall creates leaching.
- Long spells of drought, such as may occur in grassland regions, tend to increase the amount of minerals such as calcium by evaporation.

■ Topography

- Deep soils develop on level surfaces.
- Where slopes are steep thin soils develop, owing to erosion by run-off and gravity.
- Gently sloping ground helps drainage.
- High ground is exposed and is cold.
- Soils on high ground are generally thin, with few minerals.
- South-facing slopes are generally warmer and encourage growth early in the season.

■ Living Organisms

- Plant roots help to bring soil particles together, which is especially important on steep slopes.
- Plant roots absorb moisture and nutrients.
- Worms, beetles and other insects help to aerate the soil.
- Insects, fungi and bacteria break down plant matter.

■ Time

- Soils develop over long periods, often thousands of years.
- Soils can also be eroded away in short spells, only years or even months, as for example due to wind erosion in the Dust Bowl in North America.
- Soils may develop quickly or slowly, depending on the parent material.

■ Water-retention Properties

- Water in the soil is essential for plant growth.
- When water is absent plants wither and die unless they are adapted to drought conditions, as desert plants are.

- Water-retention rates are determined by soil structure and texture.
- Clay soils can hold more water than sandy soils.
- Too many clay particles lead to waterlogging.

OPTION TOPIC 33

FACTORS INFLUENCING SOIL CHARACTERISTICS

The main influencing processes are:
weathering, humification, leaching, podzolisation, laterisation, salination, calcification and gleying.

■ Weathering

- Physical weathering is the physical breakdown of rocks into smaller fragments. These fragments retain the character of the parent rock.
- Chemical weathering breaks down soil materials by carbonation, hydration, hydrolysis and oxidation.

■ Humification

This is the breakdown of plant matter into humus. It occurs most rapidly in hot, humid regions.

■ Leaching

Heavy rainfall absorbs minerals close to the surface and carries them down to lower horizons in the soil.

■ Podzolisation

- When rainwater percolates through some soils, such as acid bog or coniferous soils, it becomes increasingly acidic and removes most minerals in solution to lower horizons. This leaves the A horizon a greyish colour, with few minerals remaining there, and a reddish B horizon where minerals are deposited.
- A hard pan may develop, causing waterlogging.

■ Laterisation

- This occurs in tropical regions with high temperatures and rainfall.
- Leaching occurs, which removes all minerals except iron and aluminium.
- Concentrated deposits of both build up close to the surface.

- Aluminium deposits create bauxite and iron particles create red laterite soils.

■ Salinisation

When ground water rises to the surface by capillary attraction in hot regions, concentrations of salt occur in the surface soil layer. Here these salt deposits create a hard, toxic crust.

■ Calcification

In low-rainfall climates in the middle of continents deposits of calcium remain near the surface, due to evaporation.

■ Gleying

This occurs when surface soil is waterlogged, such as in drumlin regions. Such soils are poorly aerated, and the lack of oxygen retards decomposition.

OPTION TOPIC 34
CLASSIFICATION OF SOILS

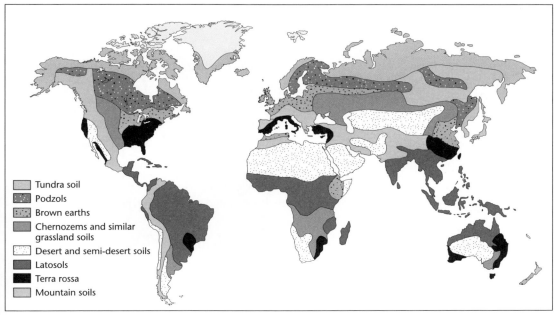

Tundra soil
Podzols
Brown earths
Chernozems and similar grassland soils
Desert and semi-desert soils
Latosols
Terra rossa
Mountain soils

▲ World zonal soil types. World soil maps do not always show the soils as they exist in reality, but instead they show the zonal soil most likely to occur in a region.

There are three basic soil groups:
Zonal, Intrazonal and Azonal.

ZONAL SOILS

Zonal soils are classified according to the climate zone in which they occur. They are mature soils with distinctive profiles and clear horizons. They include the following.

■ **Tundra Soils**

These are soils in Arctic regions such as northern Canada, Scandinavia and Russia.
- Vegetation consists of lichens, shrubs and mosses.
- They have a shallow, brown to dark-grey A horizon.
- The subsoil is permanently frozen ('permafrost').

■ **Latosols**
- These are soils of hot, humid regions within the tropics, such as India, Indonesia and Brazil.
- High rainfall has leached out most minerals except for iron and aluminium oxides.
- Iron oxides in the B horizons tint the soil a red colour, forming laterite.
- Chemical weathering is dominant and plant matter is broken down quickly.
- They are generally soils of a tropical region such as rainforest or monsoon forest.

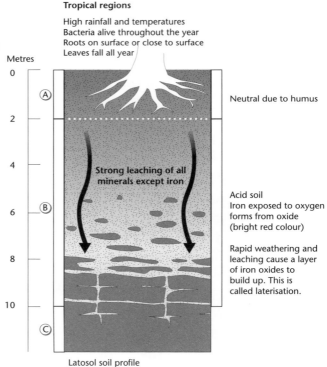

Tropical regions

High rainfall and temperatures
Bacteria alive throughout the year
Roots on surface or close to surface
Leaves fall all year

Metres

Neutral due to humus

Strong leaching of all minerals except iron

Acid soil
Iron exposed to oxygen forms from oxide
(bright red colour)

Rapid weathering and leaching cause a layer of iron oxides to build up. This is called laterisation.

Latosol soil profile

▲ Latosol soil profile

■ Podzols

- These soils occur in the coniferous forest belt in northern latitudes.
- Pine needles form the ground cover beneath the forest canopy.
- Percolating rainwater leaches out surface minerals that are redeposited to form a hard pan, which creates waterlogging.
- They are acidic and need large amounts of lime and fertiliser to make them fertile.

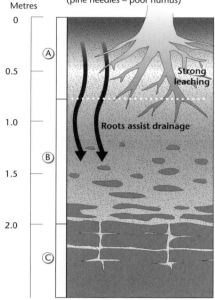

Cold regions

High latitudes or high altitudes in temperate regions
Coniferous forest
(pine needles – poor humus)

▲ Podozol soil profile

■ Chernozems (black earths)

- These are found in temperate grassland regions, such as the steppes of Russia and the Ukraine, the pampas of Argentina, the prairies of North America.
- Rainfall is low.
- The A horizon is black owing to a high humus content and little leaching.
- Calcium deposits are deposited in the B horizon.
- They are neutral soils with a crumb structure and are very fertile.

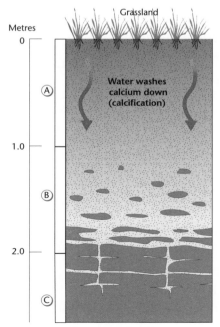

Temperate regions

Chernozem soil profile

▲ Chernozem soil profile

■ Desert Soils

- These occur in hot, dry regions of arid and semi-arid regions in temperate and tropical areas, such as the Atacama in Peru and Chile and the Californian and Nevada deserts.
- Humus is limited, owing to the almost complete absence of vegetation.
- Intense sunshine creates salt deposits from evaporation of ground moisture. This process is called **salinisation.**

INTRAZONAL SOILS

Intrazonal soils are individual soils that develop within regions of zonal soils where local factors such as parent material or drainage may have a more dominating effect than climate.

■ Peat Soils

- Peat soils are black and the surface material consists of partially decayed vegetation. The remaining soil consists of dead plant matter.
- They form in regions of cold that have persistent rainfall.
- They support only acid-loving plants such as rhododendron and heathers.
- There are two types of peat soil in Ireland:

Blanket peat

- Blanket peat covers hill and mountain tops inland and along the west-coast counties.
- Acid groundwater prevents decay of plant matter.
- It is shallow: only about 2 metres deep.

Raised peat

- Raised peat developed in shallow lakes.
- These may be 10 metres or more in depth and have been exploited commercially.

■ Gley Soils

- These form in waterlogged regions because of the presence of impermeable soils such as shales or clay.
- They have a blue-grey colour owing to lack of oxygen, and that prevents decay.
- They are found in the Cavan–Dundalk drumlin landscape, in Antrim–Derry coastal regions and in South Clare.

■ **Rendzinas**

- These form in limestone and chalk regions and their surface cover consists of grasses.
- The A horizon is black or dark brown There is no B horizon.
- The surface soil sits directly on the bedrock.
- This kind of soil is suited to beef-cattle rearing, as in the Burren or Karst regions.

■ **Terra Rossas**

- These are mature, limestone-based soils.
- Iron minerals in the soil have been oxidised, creating a red soil.
- Terra rossa soils are found in the Mezzogiorno in southern Italy and in the coffee-growing regions of Brazil.

Azonal Soils

Azonal soils are soils with an immature profile and have not had time to develop fully.

- The parent material may be weathered rock or debris that may have been transported from some other location by ice.
- Their location is not confined to any specific climatic zone.

■ **Lithosols**

- These are stony, shallow soils formed from the weathering of the bedrock.
- Erosion and sometimes mass movement may prevent the development of a soil profile.
- They are common on upland slopes.

■ **Regosols**

- These are derived from volcanic deposits, sand deposits or alluvial deposits.
- The A horizon is light in colour.
- There is no B horizon.
- The C horizon consists of silt or sand or a mixture of both.
- These soils form alluvial deposits on river floodplains, such as the Tigris–Euphrates rivers in Iraq and the Indus and Ganges flood plains in India and Pakistan.

OPTION TOPIC 35
HUMAN INTERFERENCE WITH SOIL CHARACTERISTICS

OVERCROPPING AND OVERGRAZING

Irish example: see Burren in Co. Clare above.

CASE STUDY: CAUSES OF DESERTIFICATION IN THE SAHEL

What is meant by desertification?

It refers to the reduction in vegetation cover, thereby exposing the soil to erosion by wind, rain or both, so making a region unable to provide for its natural wildlife or human populations.

■ **Desertification in North Africa**

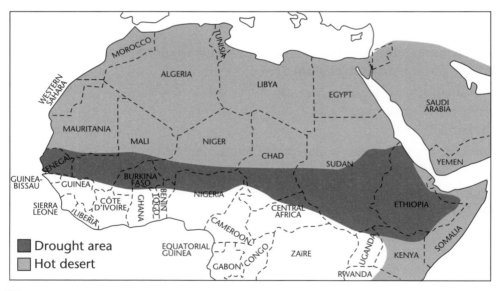

▲ Drought areas of North Africa

Climate influence

- Rainfall has reduced by at least 30 per cent over the past 10 years.
- Rain has arrived late during the wet season, so affecting wildlife.
- There have been many years of drought with below-average rainfall over the past forty years.
- Higher world temperatures are taking effect owing to global warming and increased evaporation levels.

Changing agricultural practices

- Cattle ownership indicated status or wealth, leading to increased herd numbers.
- Overgrazing was caused by increased cattle numbers in tropical grassland regions that bordered deserts or semi-desert areas.
- Increased water supply from wells led to local overgrazing and damage to the land by increased cattle numbers.
- Reduced nomadic tradition due to more intensive use of land led to overcropping and overgrazing. Land had no time to recover.
- Increased tillage for cash crops such as groundnuts on marginal land replaced grazing, so the fallow year was abandoned.
- Overcropping led to reduced yields and finally the soil became sterile.
- Reduced yields led to increased tillage area.
- Trees and bushes, such as acacia, were cut down for firewood. This led to increased wind effect and soil erosion.
- Lack of ground cover led to reduced evaporation and so less rainfall.

Soil conservation

- Contour ploughing: soil is ploughed across the slope rather than up and down the slope. Each drill ridge acts as a dam for down-slope water movement.
- Contour ploughing reduces erosion by 50 per cent.
- Terracing is best on very steep slopes that are used for tillage crops or vines.

Crop rotation

Some crops, such as nitrogen-fixing alfalfa grasses, replace the mineral content used by previous crops.

New farming methods

- Strip-cropping involves planting crops that mature at different times on widely spaced rows.
- Exposure of large regions of soil is prevented, so reducing the effects of wind erosion.
- Crops of different heights are also used for the same purpose.
- Shelter belts, such as rows of trees or shrubs, are planted to reduce wind speed.
- New animal breeds were introduced into the Sahel. These included smaller, better-quality herds that fattened more quickly or produced more milk.
- Goats and sheep were introduced to areas of poor scrub that would otherwise go unfarmed.

OPTION TOPIC 36

BIOMES

Study Temperate Deciduous Biome (pages 286–88) <u>AND</u> EITHER Desert Biome (pages 288–90) <u>OR</u> Tropical Monsoon Biome (pages 290–91)

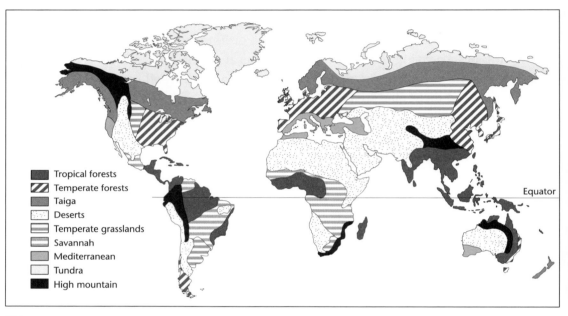

▲ The nine major biomes of the world

TEMPERATE DECIDUOUS BIOME IN EUROPE

■ **An Irish and European Example: Brown Earths: An Irish Soil Profile**

- Brown earths are found in temperate regions of deciduous forest.
- Because there is little leaching there are no distinct horizons. However, the A horizon is a little darker than others.
- They have a crumb structure and are naturally very fertile.
- Many earthworms and living organisms are found in them.

Location

Found on the western side of continents, between 40 and 60 degrees north and south of the Equator.

Influencing factors on climate

- close to mild, moist ocean conditions
- affected by warm, moist, south-westerly winds
- average winter temperatures 4° to 6° Celsius
- average summer temperatures 15° to 16° Celsius
- moisture throughout the year, with a winter maximum of 1,000–1,500 mm.

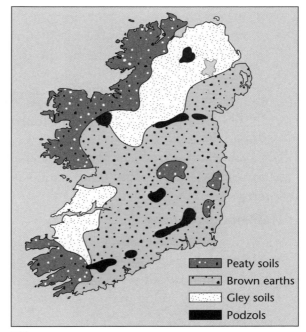

Peaty soils
Brown earths
Gley soils
Podzols

▲ Map of Ireland showing commonest soil types

Natural vegetation

- temperate deciduous forest, with ash, oak, elm, beech, chestnut, hazel, hawthorn and sycamore
- some layering
- tallest oaks, second layer ash, chestnut, elm and beech. Third layer hazel and hawthorn
- tree type influenced by soil type, as local soils vary depending on slope, bedrock or soil composition
- forest floor has many shrubs, such as ferns, mosses, brambles, orchids.

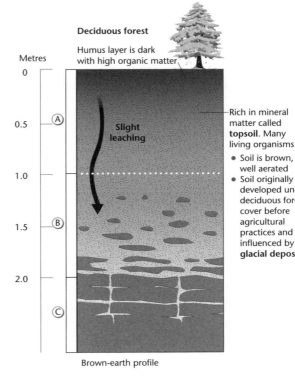

Deciduous forest

Humus layer is dark with high organic matter

Metres

Slight leaching

Rich in mineral matter called **topsoil**. Many living organisms

- Soil is brown, well aerated
- Soil originally developed under deciduous forest cover before agricultural practices and influenced by **glacial deposits**

Brown-earth profile

▲ Brown-earth profile

Soil influence
- Autumn leaf fall allows recycling of nutrients, humus is created by bacteria, and osmosis absorbs nutrients into the trees, creating new leaves.
- Earthworms mix this plant debris throughout the A horizon.
- These are deep soils, with some tree roots reaching and breaking up the bedrock.

Animal life
Grey and red squirrels, badger, fox, rodents, wild boar, wolves, rabbit, hare. Many of these have become extinct in some regions owing to human interference.

Human interference
- Most woodland has been cut down to make way for agriculture.
- Grazing of animals keeps landscape open (without trees).
- Oaks were cut down for large wooden galleons during colonisation.
- Farming has added many nutrients by means of artificial fertilisers.

Case study: Burren in Co. Clare
- Woodland cover was cut down for farming.
- Soil was tilled by early farmers and exposed to strong westerly winds.
- Soil was eroded leaving bare, rocky, limestone landscape called karst.

CHOOSE DESERT OR FOREST BIOME

DESERT BIOME

A desert region is an area characterised by little or no rainfall, where vegetation is sparse or absent.

Small local areas may have lush vegetation due to water supply available close to the surface. Such areas are called oases.

■ Desert Characteristics
- Deserts may be either hot or cold.
- Desert surfaces are generally boulders, gravels, bare rock or sand.
- All deserts have their own characteristic fauna and flora.
- Desert moisture is unpredictable in distribution and amount.
- Rainfall may occur in sudden downpours in localised areas, creating flash floods.
- Coastal mists may affect coastal deserts where cold currents exist offshore.
- Hot deserts lie between 15 and 30 degrees north and south of the Equator.
- They are affected by the trade winds, which create high-pressure zones.

- Compression of the descending air causes it to heat and retain its moisture.
- Clear blue skies and sunny weather dominate.
- Some hot deserts are due to the rain-shadow effect, for example the Atacama and the Kalahari.
- Here coastal mountains create rain on the windward side and are sufficiently high so that winds are dry as they descend on the rain-shadow side, creating drought.
- Temperate deserts lie between 30 and 40 degrees north and south of the Equator.
- Some deserts are also affected by rain-shadow and cold ocean currents, for example the Patagonian Desert in Argentina.

■ **Case Study: North American Deserts**

- The North American deserts include four regions: the Chihuahua Desert, the Sonora Desert, the Mojave Desert and the Great Basin.
- They lie between the Rocky Mountains and the Sierra Nevada mountains.
- The Mojave, Sonora and Chihuahua are hot deserts.
- The Great Basin is colder because it is more elevated than the others.

Climatic characteristics

- All suffer extremely long drought periods.
- Localised summer downpours create flash floods.
- Temperature in winter is about 8°C, while summer temperatures average 30°C.
- Diurnal range can be as great as 30°C owing to lack of cloud cover.

Soils

- Soils are aridsols.
- Soil texture varies from fine sand to gravel and stony.
- Some regions have deep soil deposits from continuous deposition of surface streams.
- Soils are poorly developed, owing to the absence of moisture for break-up of minerals and the lack of plant matter.
- Intense evaporation creates alkaline soils with calcium, sodium and gypsum minerals.
- Calcification is the dominant process.
- Salinisation also is common, creating salt deposits in saltpans.

Vegetation

- Some plants have adapted well to the short downpours. These are ephemerals that complete their life cycle in two to three weeks.
- Ephemerals open their seed pods during the downpours as a physical

reaction to the water. They sprout quickly, flower, pollinate and die within a short time.

- Some plants, such as the Giant Saguaro, store water. It has the following characteristics:
  1. needle leaves to break the wind, creating a cooling effect for the plant
  2. vertical grooves to direct water to its base, where its roots absorb it quickly
  3. waxy bark to prevent evaporation.
- Cacti are common in American deserts.

Fauna

- Desert regions have few animals owing to the lack of water supply. Those that do live in these hot, dry regions have adapted to their surroundings.
- Nocturnal or early-morning animals include the rattlesnake and the elf owl.
- The tarantula, the desert tortoise and kangaroo rat burrow into the sand to avoid the hot sunshine.
- Some animals, such as the rabbit, are dormant during the hot summer.
- Reptiles produce uric acid instead of urine, so wasting little water.
- The roadrunner, a desert bird, runs instead of flying to reduce energy loss.

OR

BIOME – TROPICAL MONSOON FOREST IN INDIA

Geographical distribution

These forests exist along the western coastal regions of India and Assam, Sri Lanka, Bangladesh, Myanmar, south-western Africa, French Guiana, north-eastern and south-eastern Brazil.

■ **Case Study: Tropical Monsoon Forest in India**

Climate

It is divided into two main seasons, the wet and dry monsoon.

Dry monsoon

- October to June; cold north-east trade winds blow from the dry continent.
- Generally dry in December, January, February and March.

Wet Monsoon

- The wet season is created by the south-west monsoon winds that blow from the Indian Ocean.
- Relief rain dominates, as the winds drop their moisture when the air rises over the Western Ghats.
- Rain falls from April to November and annual rainfall totals 2,800 mm.

- Temperature is high, averaging 27°C throughout the year. The annual range is only about 3.6°C.

Soil

- The soil is Zonal Latosols.
- High rainfall has leached out most of the minerals except iron and aluminium oxides.
- Iron oxides in the B horizon tint the soil a red colour, forming laterite.
- Chemical weathering is dominant and plant matter is broken down quickly.

Vegetation

- The monsoon forests are found on the Andaman and Nicobar Islands, the slopes of Western Ghats which fringe the Arabian coastline of peninsular India, and the greater Assam region of India in the north-east.
- Tree types include Indian rosewood and teak.
- The three-tiered forest includes the emergent layer, with giant trees up to 55 metres tall; the intermediate canopy, from 9 to 18 metres tall; and the open forest floor.
- This dense forest grades up into a more open forest where dense jungle exists at ground level.
- Brilliantly coloured epiphytes grow in the hollows of trees.
- Jungle growth, such as bamboo, is also found along streams and in the openings made by people.
- There are seven national parks along the forests of the Western Ghats.
- Many regions of forest have been felled for the lumber industry and for agriculture.
- Major threats to the forest include tea, coffee, potato, teak and eucalyptus plantations.

Fauna

- Tiger and elephant are the best known and both are threatened by fragmented forest habitat.
- Over 90 of India's 484 reptile species are endemic to these forests.
- Also endemic are 35 per cent of the plants and 75 per cent of the amphibians.

PEOPLE'S INTERFERENCE WITH BIOMES

■ Early Settlement and the Clearing of Forest Cover

- Farming, more than any other activity, has led to deforestation on a world scale.
- The knowledge of farming spread from the Middle East to Europe. It was first practised in the Middle East about 10,000 years ago.

- It reached Ireland about 6,000 years ago.
- It led to deforestation on a large scale in some regions when trees were cut down to create tillage and grow crops.

■ **Deforestation in the Burren in Co. Clare**

- The Burren had a gritty soil cover after glacial times about 10,000 years ago.
- Early farmers cut down the trees so they could till the crumbly soil.
- Over-exposure to coastal winds and rain led to erosion and loss of soil.
- Today most of the Burren is a barren, rocky landscape.

■ **The Felling of Tropical Forests**

- Large regions of tropical forest have been cut down over the past four decades.
- In Brazil this deforestation was carried out for hardwood timber supplies, such as teak and mahogany.
- Large ranches were also created in the cleared forest land to supply fast-food chains with meat.
- Soil erosion is common where deforestation has occurred.
- Plantation agriculture and wood demand in India have led to large loss of forest land.
- Native tribes lose their homeland and their way of life as they are forced from their lands.
- Big industrial projects such as hydroelectric dams lead to large-scale deforestation. Valleys are also flooded behind dams: for example, the Tucurui project in Brazil.

■ **Intensive Agricultural and Industrial Activities**

- Most agricultural lowlands that are intensively farmed were once covered by deciduous woodland.
- Such regions include Western Europe and America. Most American houses are timber-framed. This means that large quantities of wood are needed for construction.
- California's redwoods were cleared by lumber companies and miners.
- Over 500 square kilometres of natural habitat are lost to development in California each year. This includes semi-desert land around Los Angeles.
- Las Vegas has become a large, sprawling, urban region that was built in a desert environment.

OPTION 4:
THE ATMOSPHERE–OCEAN ENVIRONMENT

OPTION TOPIC 37
THE EARTH'S ATMOSPHERE

■ **Composition of the Atmosphere**

- The atmosphere is a mixture of odourless, transparent gases surrounding the Earth and held in place by the pull of gravity.
- The atmosphere has 78 per cent nitrogen, 21 per cent oxygen, and water vapour, carbon dioxide, ozone, argon, sulphur, methane, dust and pollutants making up the remaining 1 per cent.
- The highest levels of water vapour in the atmosphere occur within the tropics, owing to high evaporation rates.

■ **Structure of the Atmosphere**

- **Troposphere** (nearest the ground): 75 per cent of all the atmosphere's gases and almost all of its water vapour are concentrated in this layer. Its upper-limit height from the Earth's surface at the Equator is 16 km, and 8 km above the poles.
- **Stratosphere**: This is a cloudless zone of thin, dry air. Ozone is located in this zone. Because ozone absorbs UV radiation the temperature of the stratosphere rises with altitude.
- **Mesosphere**: Temperatures decrease with altitude to a low of about –90°C. The upper boundary line of this zone is called the mesopause.
- **Thermosphere**: Temperatures rise again, from –90°C to 1,500°C. Unusual light displays called **auroras** occur in this zone.

■ **Atmospheric Pressure**

- Warm air rises because it has a low density, causing low atmospheric pressure.

- Cold air falls because it has a high density, causing high atmospheric pressure.
- The Coriolis effect on wind movement deflects winds to their right in the northern hemisphere and to their left in the southern hemisphere.
- An area of low pressure is called a cyclone or a depression. Cyclones rotate in an anticlockwise direction in the northern hemisphere.
- An area of high pressure is called an anticyclone. Anticyclones rotate in a clockwise direction in the northern hemisphere.
- Atmospheric pressure is measured in hectopascals. Average atmospheric pressure at sea level is 1013 hPa.
- Isobars are lines joining places of equal atmospheric pressure.
- The instrument used to measure atmospheric pressure is called a barometer. There are two types of barometer, an **aneroid barometer** and a **mercury barometer**.
- An **barograph** is an aneroid barometer that produces a printout of atmospheric pressure on a graph.

■ Wind

- Wind is the movement of air from areas of high pressure to areas of low pressure.
- Local circulation is the movement of air over a distance ranging from tens of kilometres to a thousand.
- Global circulation is the movement of volumes of air in paths that carry it around the planet Earth.
- There is a constant decrease in atmospheric pressure as one flies from warmer to colder air without changing elevation.
- The rate of pressure change over a given horizontal distance is called a pressure gradient.
- A place where two surface air flows meet, so that air has to rise, is called a convergence zone. The convergence zone at latitude 60 degrees is called the Polar Front.
- Because of the Coriolis effect, circulating air in the troposphere splits into three globe-encircling convection cells in each hemisphere.
- The **Hadley cells** occur between the Equator and latitude 30 degrees.
- **Ferrel cells** are mid-latitude cells from 30 to 60 degrees.
- **Polar cells** are located between 60 degrees and the poles.

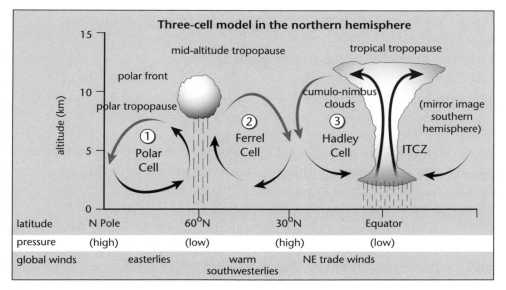

Three-cell model in the Northern Hemisphere

Solar Energy on the Earth's Surface

- Solar energy is received on Earth by short-wave radiation.
- 45 per cent of solar energy reaches the Earth by direct or indirect radiation, 26 per cent is either absorbed or reflected back into space by clouds and dust particles, 25 per cent is absorbed by the atmosphere, and 4 per cent is reflected back into space by the Earth's surface.
- The proportion of radiation reflected back into space is called the albedo.
- The albedo varies according to the ability of the surface to reflect light; over oceans it is 10 per cent, over deserts 40 per cent and from fresh snow it is 85 per cent.

Humidity

- Humidity is the percentage of water vapour in the air that it can hold at that temperature (relative humidity, RH).
- Humidity is measured by a **hygrometer.**
- Air is saturated when it contains the maximum amount of water vapour possible at that temperature, that is, its RH is 100.
- At any given temperature there is a maximum amount of water vapour that air can hold.
- Cold air can hold only small amounts of water vapour.
- Any further cooling beyond the saturation point causes condensation. This is called its dew point.

OPTION TOPIC 38
THE OCEANS

COMPOSITION OF THE OCEANS

■ Salinity

- Ocean water contains an average of 3.5 per cent dissolved salts, composed of 75 per cent halite (sodium chloride) with the remainder including gypsum, calcium sulphate and others.
- The salinity of sea water varies with location.
- Salinity reflects the balance between the addition of fresh water from rivers, or rain, and the removal of fresh water by evaporation, because when sea water evaporates salt is left behind.
- The oceans are saltier in regions with high evaporation rates and minimal rainfall.
- The saltiest oceans, such as the Mediterranean, the Red Sea and the Dead Sea, are enclosed sea areas that do not mix freely with the main oceans.
- Salinity decreases near the Equator, because of high rainfall; and in rainy, high-latitude regions; and near the mouth of large rivers.
- Surface water is saltier in warm latitudes because of evaporation and is less salty at high latitudes because of melting ice.
- The salt content of deep ocean water below 1 km is even at all latitudes.
- The halocline is the boundary between surface-water salinity and deep-water salinity.

■ Temperature of the Oceans

- Surface water is hottest near the Equator and coldest at the poles.
- Average global sea-surface temperature is about 17°C, but it is near freezing close to the poles and rises to 35°C in restricted tropical seas.
- Sea-surface temperatures also vary with the seasons. The difference is about 2°C in the tropics, 8°C in temperate latitudes and 4°C near the poles.
- Surface sea water is warmest, and the deep sea is coldest.
- The oceans may be divided into layers:
  1. the sunlight zone, from 0 to 200 m,
  2. the twilight zone, from 200 to 1,000 m,
  3. the dark zone, from 1,000 to 6,000 m.
  4. Below 6,000 m there is no light at all.

■ Ocean Water Density and Pressure

- Salt content and temperature affect density.
- Density increases with salinity.
- Density increases:
  - when water is cooled,
  - when evaporation increases salinity or
  - when saline water freezes.
- Dense water tends to sink.
- The deeper the water, the greater the pressure due to the weight of overlying water.

■ Ocean Currents

1. Ocean currents that create large circular flow patterns are called **gyres.**
2. Gyres rotate clockwise in the northern hemisphere and anticlockwise in the southern hemisphere. This is called the **Coriolis effect.**
3. Sailors refer to the centre of the North Atlantic Gyre as the **Sargasso Sea** because its non-circulating waters accumulate sargassum, a tropical seaweed.
4. **Downwelling zones** are places where near-surface water sinks; **upwelling zones** are places where deep water rises.
5. Surface water sinks where winds and the **Coriolis effect** carry surface water towards a coast, because there is an oversupply.
6. If the Coriolis effect moves water away from a coast, there is a deficit of water near the coast so deep water rises.
7. The rising and sinking of sea water owing to temperature and density differences is called **thermohaline circulation**.
8. During thermohaline circulation cold, salty water tends to sink and warm, less salty water rises.
9. So the cold water in polar regions sinks and flows along the bottom of the ocean towards the Equator.
10. The circulation of ocean waters involves surface currents as well as movements of deep-water masses.
11. Surface currents are created by wind.
12. Currents that flow from lower latitudes to higher latitudes are called warm currents.
13. Currents that flow from higher latitudes to lower latitudes are called cold currents.

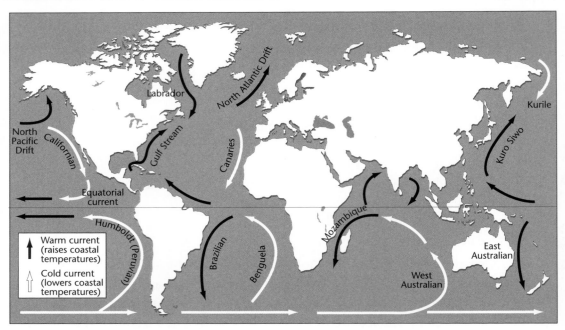

▲ Major ocean currents

FACTORS THAT AFFECT OCEAN CURRENTS

1. **Winds:** These are the primary force affecting ocean currents. For example, the Monsoon winds in the Indian Ocean affect the direction of the ocean currents: when the winds are from the south-west, so too is the direction of the ocean drift.

2. **The Coriolis force:** The directions of ocean currents are altered by the rotation of the Earth on its axis. Ocean currents are deflected to the right in the northern hemisphere and to the left in the southern hemisphere.

3. **Density differences** also affect ocean currents.

■ The Effects of Ocean Currents

- Ocean currents help to distribute heat from warm regions to cold regions.
- The south-west anti-trade winds are warmed as they pass over the North Atlantic Drift and so help to keep Western Europe mostly ice-free during winter.
- Air masses that pass over warm oceans absorb moisture and carry it inland, so helping farming and water supplies for people.
- In some cases cold currents cause condensation over sea areas, so creating dry winds over land and leading to deserts like the Atacama in Chile and Peru.

- Cold currents bring cold weather and cold climates to places in middle latitudes. For example the **Kamchatka** in the Pacific brings freezing weather to the north-west Pacific coastline in winter, as does the **Labrador current** to the north-west Atlantic coastline.
- Cold currents contribute to the formation of deserts; for example:
  - The Peru current affects the Atacama.
  - The Benguela Current affects the Kalahari desert.

OPTION TOPIC 39

MOISTURE IN THE ATMOSPHERE

THE WATER CYCLE

- Evaporation occurs mostly over the oceans and falls mostly over the land.
- Water vapour rises over the oceans and forms clouds. The clouds are blown inland where they drop their moisture in the form of rain, hail, sleet, snow or dew.
- Coastal mountain ranges create relief rain if the region experiences onshore winds.
- Rainfall over land regions seeps into the ground and finally into streams, which carry it back to the sea to complete the water cycle.
- Some rainfall over land is returned directly back into the atmosphere by the processes of transpiration and evaporation.

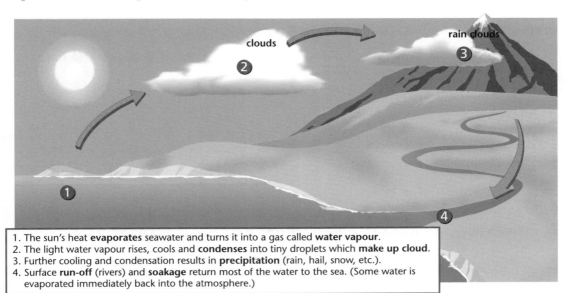

1. The sun's heat **evaporates** seawater and turns it into a gas called **water vapour**.
2. The light water vapour rises, cools and **condenses** into tiny droplets which **make up cloud**.
3. Further cooling and condensation results in **precipitation** (rain, hail, snow, etc.).
4. Surface **run-off** (rivers) and **soakage** return most of the water to the sea. (Some water is evaporated immediately back into the atmosphere.)

▲ The water cycle

■ Forms of Condensation

- Dew occurs at night when the cold ground cools warm air above it.
- Frost is frozen water droplets created when ground temperatures are below zero.
- Fog occurs when air is close to saturation point. Tiny water droplets remain suspended in the air and these reduce visibility.
- There are two types of fog: advection fog and radiation fog.

■ Types of Cloud

- Cirrus are high, wispy clouds.
- Stratus form when clouds join to form a layer, altostratus are high stratus associated with depressions, and nimbostratus are dark, rain-bearing clouds.
- Cumulus are fluffy clouds.
- Cumulonimbus are very dark, anvil-shaped clouds associated with thunderstorms.

OPTION TOPIC 40

RAINFALL

There are three main types of rainfall.

RELIEF RAIN

Occurs regularly in the west of Ireland, e.g. in the mountains of Mayo and Kerry.

1. Sea winds are laden with moisture.
2. They are forced to rise over mountains.
3. As they rise, they are cooled.
4. Water vapour condenses and falls as rain on the windward side of the mountains.
5. The leeward or sheltered side of the mountain gets little rain because it is in the rain shadow. As air descends on the leeward side it becomes warmer, is able to hold moisture and so is dry.
6. Many of the wettest places on Earth are on the windward side of mountains.
7. Rainfall increases where mountain ranges lie parallel to the coastline.

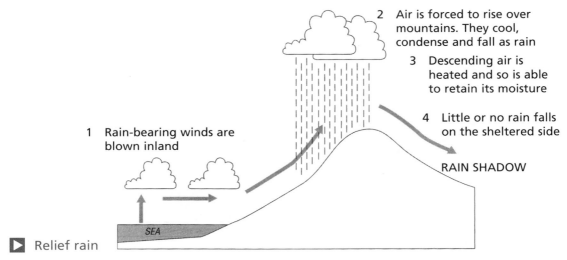

2 Air is forced to rise over mountains. They cool, condense and fall as rain

3 Descending air is heated and so is able to retain its moisture

1 Rain-bearing winds are blown inland

4 Little or no rain falls on the sheltered side

RAIN SHADOW

SEA

▶ Relief rain

CONVECTIONAL RAIN

This occurs regularly at the Equator, and also throughout Ireland in summer.

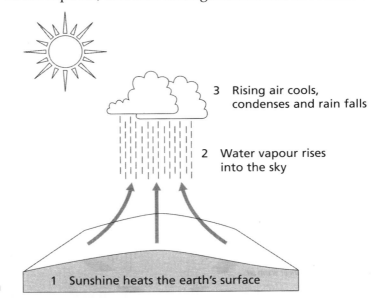

3 Rising air cools, condenses and rain falls

2 Water vapour rises into the sky

1 Sunshine heats the earth's surface

▶ Convectional rain

CYCLONIC RAIN

Occurs regularly in the North Atlantic and Ireland, especially in winter.

1. Cold, polar air meets warm, tropical air over the North Atlantic.
2. Cold air cuts in under the warm air and forces it to rise quickly.
3. As the warm air rises, water vapour condenses and forms cloud along both the warm and cold fronts.
4. Rain falls as showers, giving rise to changeable weather.
5. Cyclones blow across the North Atlantic in a north-easterly direction.

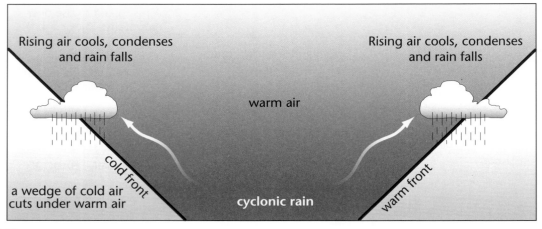

Rising air cools, condenses and rain falls

Rising air cools, condenses and rain falls

warm air

cold front

a wedge of cold air cuts under warm air

cyclonic rain

warm front

▲ Cyclonic rain

OPTION TOPIC 41
WIND

THE CORIOLIS EFFECT

- The rotation of the Earth causes winds and ocean currents to be deflected to the right in the Northern Hemisphere and to the left in the Southern Hemisphere.

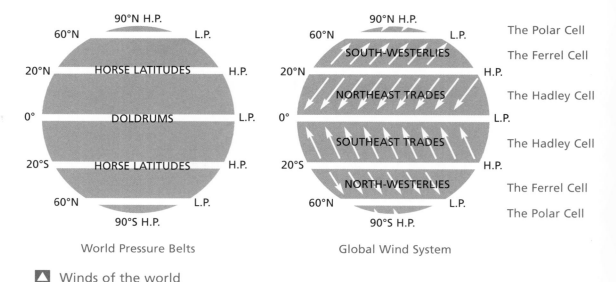

World Pressure Belts

Global Wind System

▲ Winds of the world

■ World Distribution of Temperature

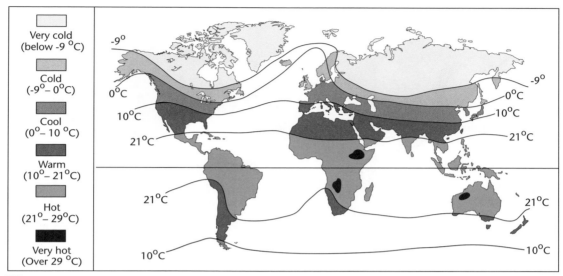

Very cold
(below -9 °C)

Cold
(-9°– 0°C)

Cool
(0°– 10 °C)

Warm
(10°– 21°C)

Hot
(21°– 29°C)

Very hot
(Over 29 °C)

▲ January temperatures in °C

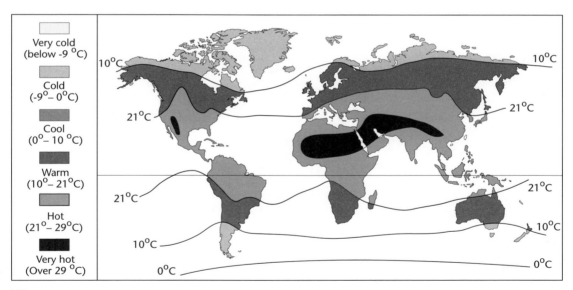

Very cold
(below -9 °C)

Cold
(-9°– 0°C)

Cool
(0°– 10 °C)

Warm
(10°– 21°C)

Hot
(21°– 29°C)

Very hot
(Over 29 °C)

▲ July temperatures in °C

■ World Distribution of Precipitation

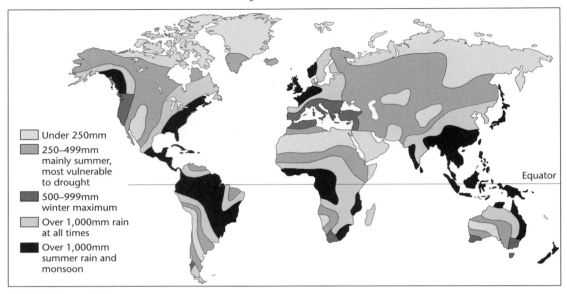

Under 250mm

250–499mm
mainly summer,
most vulnerable
to drought

500–999mm
winter maximum

Over 1,000mm rain
at all times

Over 1,000mm
summer rain and
monsoon

Equator

▲ World precipitation – total amounts and seasonal distribution

■ Land and Sea Breezes

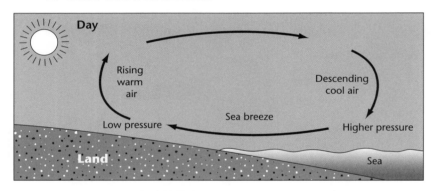

Day

Rising
warm
air

Descending
cool air

Low pressure ← Sea breeze → Higher pressure

Land

Sea

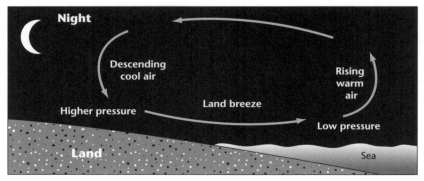

Night

Descending
cool air

Rising
warm
air

Higher pressure → Land breeze → Low pressure

Land

Sea

▲ Land and sea breezes, by day and by night

- When a region is hotter than a neighbouring region, air moves into the hot region from the cooler region. This happens so that cool air can replace the hot air that has expanded and risen.
- The air which moves in is a wind.
- By day land becomes warmer than the sea, and so air pressure is lower over land than over sea.
- Air then blows from sea to land as a breeze.
- At night the land cools more quickly than the sea, and so air blows from land to sea.

■ Local Winds

Chinook

The Chinook is a warm wind that melts snow from the Prairies and allows the early planting of wheat.

Santa Ana

The Santa Ana is a dry, strong wind that blows from the Mojave Desert and regularly fuels bush fires.

The Föhn

The Föhn is a strong, warm, dry wind that heats up as it descends from the Alps. It may cause temperatures to rise significantly and leads to avalanches.

The Mistral

The Mistral is a strong, cold, dry wind from the Alps. It sometimes destroys fruit blossoms and devastates early crops.

Sirocco

The Sirocco is a warm wind from the Sahara. It often carries red desert dust northwards, sometimes as far as Ireland.

▲ Local winds in the Americas

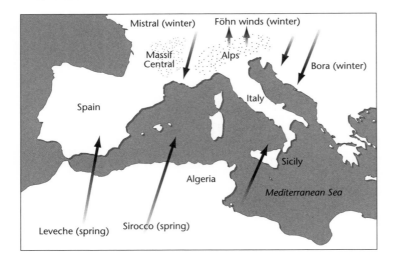

◀ Local winds across Europe

Some Other Winds

Mountain and valley winds

- The unequal heating of higher mountain slopes and valley floors creates an up-valley wind called an anabatic wind.
- In the afternoon the valley floors become warmer, so cool, denser air from upper slopes moves down the valley as a katabatic.

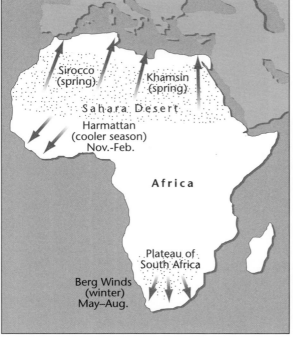

◀ Other winds in Africa

▼ Other winds in Australia

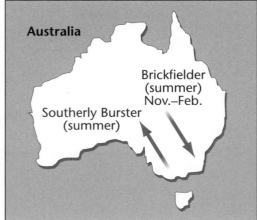

OPTION TOPIC 42
WORLD CLIMATES

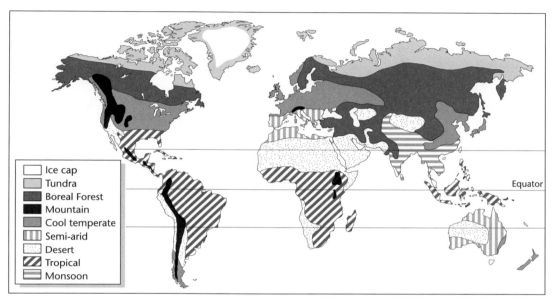

▲ Global climates

STUDY ONE GLOBAL CLIMATE:
EITHER EQUATORIAL OR MONSOON

■ Main Characteristics of Equatorial Climate

Location: 5 degrees north to 5 south of the Equator.

Temperature: Hot throughout the year; daily average 27–28°C. Sun is directly overhead or at a very high angle. One season only. Always hot, with growth throughout the year.

Rainfall: 2,000 mm. Rain falls daily in heavy thunder showers. Mostly convectional rain, high humidity.

12 hours day and night throughout the year.

Annual temperature range very small, only 1–2°C.

Pressure: Low pressure due to high evaporation rates.

OR

■ **Main Characteristics of Tropical Continental Monsoon**

Location: South Asia, South-East Asia, northern Australia.

Temperature: There are three seasons:

(a) **The cool, dry season, mid-December to end of February**
Breezes blow from land to sea. India is sheltered by the Himalayas from cold continental winds. Temperatures range from 15°C in the north to 30°C closer to the Equator. Rainfall is very light.

(b) **The hot, dry season, March to end of May**
Early winds blow from the land. Dust storms are common. Later in May southerly sea winds bring some moisture. Temperatures range from 26°C to 40°C.

(c) **Hot, wet season, from June to December**
Winds blow from the warm tropical seas inland. Temperatures can reach as high as 40°C or more. Moisture-laden sea

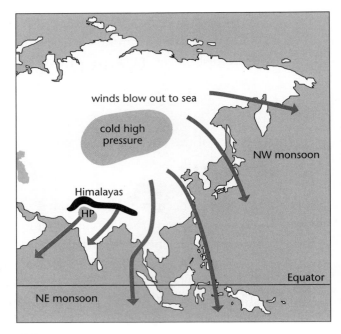

▲ Dry Monsoon

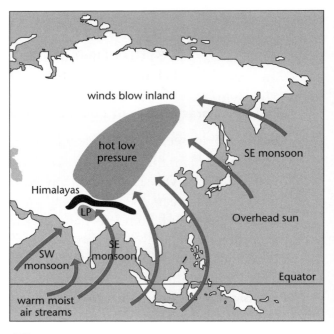

▲ Wet Monsoon

winds blow into the low-pressure cell that is located over north-east India. Heavy, continuous rainfall occurs. Cherrapunji, in the Kasi Hills, is one of the wettest places in the world with rainfall in excess of 10,000 mm, and up to 20,000 mm has been recorded.

■ Climate Change

The climates of the world have changed over time in the past and they continue to do so today.

Ice ages

- Ice ages have occurred due to changes in the orbital motion of the Earth.
- The coldest periods may have occurred when three variations in the Earth's orbit coincided. These variations are:
    1. Every 41,000 years, the tilt in the Earth's axis is more vertical than usual.
    2. Every 100,000 years the orbit of the Earth becomes more elliptical in shape.
    3. The Earth's axis 'wobbles' and describes a circle every 23,000 years, causing the sun's rays to strike at a lower angle on the northern hemisphere.
- During the ice ages the difference between summer and winter temperatures is smaller, so that snow accumulations build up over time causing ice sheets to cover large regions of the globe.

Global warming

- Global warming refers to the build-up of excess heat in the atmosphere due to an imbalance in percentages of greenhouse gases caused by people's lifestyles.
- The increase is mainly due to the increase in the burning of fossil fuels as the world's population rises and Western lifestyles spread.
- There is a large increase in gases such as methane, produced by increased livestock numbers.
- The slash-and-burn practices of tropical forest regions also create increased levels of gases such as carbon dioxide.
- Greater and greater amounts of heat radiated from Earth are being retained in the atmosphere by these increased greenhouse gases.

El Niño

- El Niño refers to a build-up of warm water off the coast of Peru for a longer period than normal.
- Usually this occurs for only a short period, followed by a long period of cold water from the Peru current.

- The cold waters of the Peru current help to bring cold waters rich in nutrients from the ocean depths to the surface. When El Niño's warm waters remain for up to a year, this pattern changes and destroys fish stocks off Ecuador and Peru.
- El Niño reverses the normally dry, desert conditions of the Atacama. Heavy torrential downpours occur, often leading to landslides and mudslides.
- Other regions such as Indonesia, which are normally wet throughout the year, suffer extreme drought leading to bush fires.

CLIMATE CHARACTERISTICS INFLUENCE ECONOMIC DEVELOPMENT

■ Heavy Rainfall in the West of Ireland

- The mountains of the West of Ireland create heavy rainfall in the region.
- Over 2,000 mm of rain falls over parts of this region.
- The heavy rainfall leaches minerals from the soil, creating podzols.
- Blanket bog develops in the wet upland and lowland regions in the West.
- The heavy rains and leached soils create a poor farming environment, restricting economic development.

■ Low Rainfall Levels in the East of Ireland

- Rainfall totals only 800 to 1,000 mm in the east.
- Soils are well-drained, rich, brown soils high in mineral content.
- Agricultural output is high.

■ Domestic Water Supplies

- Ireland's heavy and well-distributed rainfall provides plenty of water for domestic activities.
- Some regions of the world, such as the Thar Desert in northern India, experience extreme water shortages.
- Similar water shortages are experienced in Pakistan.
- The extraction of water for agricultural purposes and domestic supplies regularly leads to extremely low levels of water in some rivers in Pakistan and near the Aral Sea (see page 265).
- California's Central Valley uses irrigation water brought by canal from the Sacramento Valley.

■ The Impact of Drought and Desertification on Economic Development

See Desertification, pages 215–17.

■ The Impact of Climate on Tourism

- The Mediterranean coastline has become one of the world's greatest holiday regions because of the characteristics of its climate.
- Hot, dry summers and mild, warm winters attract tourists all year round. Summer is high season, with over 100 million tourists visiting the area.
- Temperatures of 28°C to 35°C are guaranteed throughout June, July, August and September.
- Warm winters with temperatures of 15°C on average attract many retired people to the Mediterranean.
- Longer sun holidays can be experienced by Europeans of more northerly countries with the purchase of holiday homes in countries such as Spain.
- Winter skiing in Alpine regions has grown dramatically over recent decades. Cheap air fares have made such trips an annual event for many people.

STATE EXAMINATIONS COMMISSION

LEAVING CERTIFICATE GEOGRAPHY
NEW SYLLABUS
EXEMPLAR QUESTIONS

N.B. These Exemplars are for information and consultation purposes only.

EXEMPLAR QUESTION 1 – CORE – HIGHER LEVEL

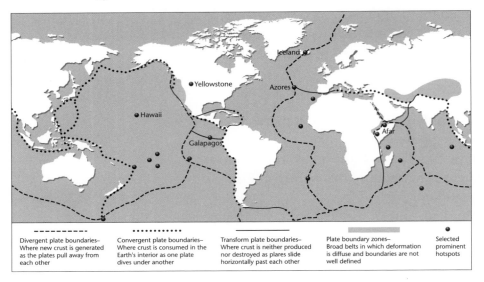

Examine the map of global crustal plates above.

A. Name **two** converging plates **and two** diverging plates.   [20 marks]

B. With the aid of a labelled diagram, explain the process of plate
    movement.   [30 marks]

C. With reference to the map, examine the relationship between the
    Earth's plates and volcanic activity.   [30 marks]

*N.B. In a question set at Ordinary Level, the crustal plates would be named, 2 margins
marked and arrows would be used to indicate plate movement. Thus, part A would be:*

•   *Name the type of plate margin at X and at Y.*
•   *Name two of the crustal hot spots shown on the map.*

EXEMPLAR QUESTION 2 – CORE – HIGHER LEVEL

Study the Ordnance Survey 1: 50000 map extract supplied.

A. From the map, identify – using six-figure grid references – an example of **each** of the following features along the course of the River Moy:
- a meander
- a confluence with another named river
- a wide flood-plain
- a levee.   [20 marks]

B. The present-day landscape of Ireland is the product of a wide range of surface processes.
Examine this statement, with detailed reference to **one** landform which you have studied. Illustrate your answer with a clearly labelled diagram/diagrams.   [30 marks]

C. Human activities can interfere with the operation of surface processes. Examine in detail **one** example of such interference.   [30 marks]

EXEMPLAR QUESTION 3 – CORE – HIGHER LEVEL

A. Draw a sketch map of Ireland. On it mark and name **two** contrasting regions.   [20 marks]

B. Explain how (i) relief and (ii) climate have influenced the development of agriculture in a non-Irish European region which you have studied.   [30 marks]

C. Examine in detail **one** of the economic challenges facing a non-European continental/subcontinental region which you have studied.   [30 marks]

EXEMPLAR QUESTION 4 – ELECTIVE – HIGHER LEVEL

A. Study this pie-graph showing how the retail price of a jar of coffee is shared among the various groups involved in the coffee trade.
- Calculate in Euro the share which the retailers receive.
- Which group receives the lowest share?
- Calculate the percentage share which the Shippers and Roasters receive.   [20 marks]

Growers €0.36

Retailers

Exporters €0.44

Shippers and Roasters €2.20

B. Explain why the producers of commodities such as coffee receive so little of the final retail price.   [30 marks]

C. Ireland has experienced both success and failure as a location for Multinational Companies.

Explain **one** reason why Ireland continues to be a favoured location for MNCs **and also one** reason why some MNCs have left Ireland to move to other locations.   [30 marks]

EXEMPLAR QUESTION 5 – ELECTIVE – HIGHER LEVEL

A. Study this chart, showing Irish Migration Patterns between 1987 and 2004.

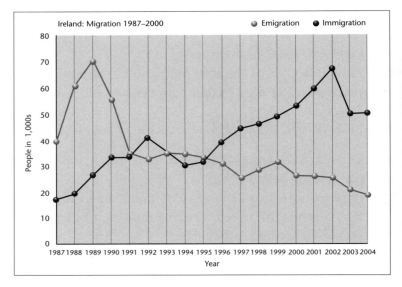

•   In which year did emigration peak?

•   When did immigration first exceed emigration?

•   Calculate the difference between emigration and immigration in 2003.   [20 marks]

B. Study the 1: 5000 Ordnance Survey extract supplied with this paper.

1) Name and locate, using six-figure grid references, **two** patterns which can be identified in the rural settlement of the region shown on the extract.

2) Explain why **each** of the patterns developed.   [30 marks]

C. Explain how the unwise development of a natural resource has created overpopulation in a region which you have studied.   [30 marks]